GEOLOGY AND WATER

DEVELOPMENTS IN APPLIED EARTH SCIENCES

VOLUME 1

GEOLOGY AND WATER

An introduction to fluid mechanics for geologists

by

RICHARD E. CHAPMAN

University of Queensland,
St. Lucia, Brisbane, Australia

MARTINUS NIJHOFF / DR. W. JUNK PUBLISHERS / 1981

THE HAGUE / BOSTON / LONDON

Distributors:

for the United States and Canada

Kluwer Boston, Inc.
190 Old Derby Street
Hingham, MA 02043
USA

for all other countries

Kluwer Academic Publishers Group
Distribution Center
P.O. Box 322
3300 AH Dordrecht
The Netherlands

This volume is listed in the Library of Congress Cataloging in Publication Data

ISBN 90-247-2455-4 (this volume)
ISBN 90-247-2437-6 (series)

PRINTED IN THE NETHERLANDS

THIS BOOK
is dedicated
to my wife
JUNE
with love and affection

Where is the Life we have lost in living?
Where is the wisdom we have lost in knowledge?
Where is the knowledge we have lost in information?
T.S. Eliot, 1934, *The Rock*

CONTENTS

PREFACE

Water is one of the world's threatened resources: it is also a substance of importance in Geology. For some years I have felt the need for a book that sets out the fundamentals of fluid mechanics, written for geologists rather than engineers. The efforts to repair my own deficiencies in this respect led me along various unfamiliar paths, few of which were unrewarding. This book is the result of my journeys through the literature and as a geologist in several parts of the world.

It has been written for students of geology of all ages, in the simplest terms possible, and it has one objective: to provide a *basis* for an understanding of the mechanical role of water in geology. It has not been written for experts in ground-water hydrology, or specialists in the fluid aspects of structural geology: it has been written for geologists like me who are not very good mathematicians, so that we can take water better into account in our normal geological work, whatever it might be. The fundamentals apply equally to mineralization, geochemistry, and vulcanology although they have not been specifically mentioned. It has also been written for the university student of geology so that he or she may start a career with some appreciation of the importance of water, and understanding of its movement.

A casual glance through the references at the end of each chapter (which include works of interest in addition to those specifically cited) will reveal many works published over 50 years ago, some over 100, and a few over 200 years ago. This should not come as a surprise, because there has been little new at an elementary level in the last 40 years. The very old references have been included because many of them reveal deep insight into the phenomena they discuss, and I have found second-hand reports of their contents frequently misleading. I hope the reader will find them as rewarding as I have. In any case, let the reader not underestimate the value of the older work. For example, Poiseuille's expression for the kinematic viscosity of water at temperatures to 45 °C (the range of his experiments), published in 1841, could be used today with little error.

Stokes would doubtless be appalled at some of the uses to which the law that bears his name is put, because he realized its limitations. Perrault is sometimes called 'the father of ground-water hydrology', but Mariotte is more worthy of that title. There is little doubt that Perrault failed to understand most of what he discussed, while Mariotte had a very good understanding of his subjects.

From around the turn of the Century, the works of King and of Veatch are full of interest; and I suspect that the pace of modern development would preclude repetition of many of their studies of the fluctuations of water-levels in wells.

Perhaps the greatest surprise of my journeys through the literature concerns the origin of what is now known as the Ghijben-Herzberg relationship, in coastal aquifers. Badon Ghijben, to use his proper name, was the *junior* author of the note that has so often been cited (one wonders how many of those who cited him had actually seen the note). Drabbe and Badon Ghijben clearly understood the movement of ground water under their dunes, but they are only remembered for the 1:42 ratio of the Ghijben-Herzberg lens. And I find it hard to believe that Herzberg, writing 12 years later, had not heard of Drabbe and Badon Ghijben from just along the coast into Holland. Soon afterwards, another outstanding Dutchman appeared – Jan Versluys – whose work has been almost entirely neglected. True, he wrote much of his early work in Dutch (but so did Drabbe and Badon Ghijben), and his paper on the dune-water theory, so full of understanding and perception, was incompletely translated into English and published without his diagrams.

In drawing attention to these early workers, it has not been my intention to denigrate later workers – far from it. Outstanding amongst later workers in the context of geology is M. King Hubbert, alone and with various collaborators. His contributions to the understanding of fluid flow, and his applications of mechanics and fluid mechanics to structural geology, have been central to my studies of the last decade or so. Forty years ago he pointed out a number of common misunderstandings in the matter of fluid flow. But the misunderstandings have persisted, and are certainly no less common now than they were then.

I have sought understanding at a simpler level, and in so doing make no apologies. The simple approach has positive merits, not the least of which is that it can be understood by people with few mathematical skills. A simple approach yields an expression for intrinsic permeability that takes porosity and the tortuosity of the pore passages into account – at least to the point where an understanding of their roles is achieved. It shows Darcy's law to be applicable to porous glass as well as to glass beads, and that the dimensionless coefficients in these two extremes differ only by a factor of about two, whereas the permeabilities differ by a factor of at least 10^9. This in turn suggests that fluid movement in very small pores is no different in principle from that in large pores (in spite of many published statements to the contrary). An extension of this simple approach to the vexed question of the flow of two immiscible liquids through porous material yields results that are consistent with experimental data without calling on anisotropy or any other feature to account for the shape of relative permeability curves, and the well-known fact that the sum of the relative permeabilities of two immiscible liquids is always less than unity.

In structural geology, the simple approach leads to results of the correct order of magnitude for such things as the proportion of a block that can be overthrust by sliding on a gentle slope.

It is my hope that readers will not only find these and the other topics of interest, but will use them as a stepping-stone to better, more rigorous, studies.

The first two chapters outline the fundamentals of fluid statics and dynamics, and develop simple equations for flow in open channels and in pipes. The third chapter builds on this base, and examines Darcy's law and the flow of liquids through porous materials. The coefficient of permeability is expanded to include the relevant measurable properties of sedimentary rocks, and the upper and lower limits of Darcy's law are examined.

In Chapter 4 we take a larger view, looking at the flow of water through aquifers and comparing the influence of a producing water-well with the influence of a stream or river. This is followed in Chapter 5 by some examples of the interaction between aquifers, springs and rivers.

Chapter 6 examines the important topic of abnormally high pore pressures in sediments, and their probable and possible causes. Interest in this topic should not be confined to petroleum geologists: it is a common phenomenon of great consequence in the early deformation of many sedimentary basins – which is the topic of Chapter 7. This leads on, in the next chapter, to another important role of water in geology: the sliding of large blocks of sedimentary rocks. Both lubricated and unlubricated sliding are considered.

In Chapter 9 we conclude with two examples of the application of the principles discussed in early chapters: a qualitative discussion of water flow along fault-planes, and a semi-quantitative analysis of relative permeability when a porous material is saturated with two immiscible liquids.

This work has, of course, involved the study of many works. In the Appendix I have reviewed a number of papers that are commonly quoted in the geological literature (and translated their equations into the notation of this book).

SI units have been used throughout. They are, I believe, a great help once one gets used to them. All students are brought up with them in school and university: the older geologists will have to conform. But I have not used all the prefixes. Conversion tables giving the SI units in various other measures are included. A simple pocket calculator will enable the reader to make cross-conversions between non-SI units.

A Glossary is also included. I have had great difficulty with this because many accepted definitions in ground-water hydrology are inconsistent with the physical principles that are involved. For example, it makes *no sense* to define 'hydraulic gradient' as the rate of change of *pressure head* because pressure head changes in an inclined aquifer in which the water is static.

ACKNOWLEDGEMENTS

During the preparation of this book I have received help, advice, and constructive criticism from many people whose expertise lies in the fields in which I was seeking to repair my deficiencies. I am conscious of having discussed some aspect of this work with almost every chemical engineer, civil engineer, applied mathematician, physicist, geophysicist and geologist in this University. To them all, I am deeply grateful. I have also benefitted from the advice of others around the world. In particular, I express my thanks to C.J. Apelt, J.H. Doveton, J.M. Fitz-Gerald, R.G. Font, V.G. Hart, D.A.L. Jenkins, W. Käss, R.A. Nelson, N. Street, and J.P. Webb.

This work would still be incomplete were it not for the tenacity, tolerance and good humour of Carolyn Willadsen (in the early stages) then Margaret Eva in the Geology Library of this University. They tracked down incorrect and unverifiable references, and acquired numerous works from other libraries. I am most grateful to them, and to their assistants.

I am also very grateful to Mrs Irene Lenneberg for drawing almost all the illustrations.

For data, and permission to publish their previously-published figures, I am indebted to R.A. Downing for Figure 5-4, W. Käss for Figures 5-5 and 5-6, K. Lemcke for Figure 5-7, and M.A. Habermehl for Figures 5-17 and 5-18.

I acknowledge with gratitude the following for permission to include data they have provided and material of which they hold the copyright:

The American Association of Petroleum Geologists, Tulsa, for Figures 1-17, 7-5 and 8-4;

The British Petroleum Company, London, for Figures 7-8, 7-10, and 8-9;

The Director of the Bureau of Mineral Resources, Canberra, for Figures 5-17 and 5-18;

Elsevier Scientific Publishing Company for several figures used in my book *Petroleum Geology.*

The Institution of Water Engineers and Scientists, London, for Figure 5-4;

The Royal Dutch/Shell Group for the data of Figures 6-3 and 6-12;

The Texas Department of Water Resources for Figures 1-16 and 5-2.

If I have inadvertently omitted any name, I apologize. All weaknesses of the book are due to me, and not to those who have done so much to ensure that it is free of error.

Brisbane, 29 June 1980

The following work, received when this book was in press, will be of interest to readers of Russian:

Gurevich, A.E., 1980. *Prakticheskoe rukovodstvo po izucheniyu dvizheniya podzemiȳkh vod pri poiskakh poleziȳkh iskopaemȳkh.* Nedra, Leningrad, 216 pp.

SYMBOLS

A	Area $[L^2]$
C, c	constants
C_f	drag or resistance, coefficient, dimensionless
D	diameter available to fluid flow $[L]$
d	diameter of granular material $[L]$
d_*	hydraulic equivalent size of quartz sphere $[L]$
E	energy $[ML^2T^{-2}]$
e	base of natural or Napierian logarithms
F	force $[MLT^{-2}]$. Formation Resistivity Factor, dimensionless
f	fractional porosity, dimensionless
Fr	Froude number, dimensionless
g	acceleration due to gravity $[LT^{-2}]$
h	head, thickness $[L]$
K	coefficient of permeability, hydraulic conductivity $[LT^{-1}]$
k	intrinsic permeability $[L^2]$
k_o	effective permeability to oil $[L^2]$
$k_{ro} = k_o/k$	relative permeability to oil, dimensionless
L	dimension of length
l	length $[L]$
M	dimension of mass
m	'cementation' factor, dimensionless
N	newton, SI unit of force (kg m s^{-2})
Pa	pascal, SI unit of pressure (N m^{-2} = kg m^{-1}s^{-2})
p	pressure $[ML^{-1}T^{-2}]$
p_e	pressure of ambient fluid
Q	volumetric flow rate $[L^3T^{-1}]$
$q = Q/A$	specific discharge, or discharge velocity $[LT^{-1}]$
R	hydraulic radius $[L]$
r	radius, correlation coefficient
Re	Reynolds number, dimensionless
S	specific surface $[L^{-1}]$, hydraulic gradient (dimensionless), total normal stress $[ML^{-1}T^{-2}]$

s	saturation, proportion of pore space occupied by fluid, dimensionless (esp., water saturation)
T	dimension of time, tangential component of total stress $[ML^{-1}T^{-2}]$, tortuosity, dimensionless
t	thickness $[L]$, time $[T]$
$U = gz$	gravity potential $[L^2T^{-2}]$
V	velocity $[LT^{-1}]$
v	volume $[L^3]$
W, w	weight, partial weight $[MLT^{-2}]$, width $[L]$
x	exponent in porosity term in intrinsic permeability (dimensionless), horizontal co-ordinate $[L]$
y	horizontal co-ordinate $[L]$
z	vertical co-ordinate $[L]$
α, β	Angles, degrees (dimensionless)
γ	weight density, specific weight $[ML^{-2}T^{-2}]$
Δ	difference of, e.g., Δh, Δt
δ	dimensionless parameter that takes pore-fluid and ambient-fluid pressures into account in various mechanical contexts; ratio of effective normal stress to active normal stress, σ/σ_e
ε	void ratio $= f/(1-f)$, dimensionless
η	dynamic viscosity $[ML^{-1}T^{-1}]$
θ	angle of slope, degrees, dimensionless
$\lambda = p/S$	proportion of overburden supported by pore fluid, dimensionless
$\lambda_e = p_e/S$	proportion of overburden supported by ambient fluid
μ	micro (10^{-6}) μs, micro-second
$\nu = \eta/\rho$	kinematic viscosity $[L^2T^{-1}]$
Π	dimensionless numbers
ρ	mass density $[ML^{-3}]$
σ	effective normal stress $[ML^{-1}T^{-2}]$, surface tension $[MT^{-2}]$
σ_e	active normal stress, in absence of abnormal pore-fluid pressures is equal to effective normal stress
T	tangential component of total stress $[ML^{-1}T^{-2}]$
τ	frictional shear stress parallel to sliding surface $[ML^{-1}T^{-2}]$
τ_0	cohesive strength, shear strength when $\sigma = 0$ $[ML^{-1}T^{-2}]$
Φ	fluid potential (Hubbert's) $[L^2T^{-2}]$
ϕ	angle of friction, grade scale
ψ	function of...

1. INTRODUCTION: LIQUIDS AT REST

INTRODUCTION

Geology is the study of rocks, and we tend to think of rocks in terms of the solid mineral constituents. This is natural because almost the rocks we examine, in outcrop or in hand speciment, are dry. We are apt to forget that, at depths below only a few metres from the surface, almost all rocks in nature are saturated with water (exceptionally and locally, also oil or gas). Some of this water is important to us as a source of fresh water for drinking, agriculture and industry: much of it is not, being too salty. But whether we can use it or not, it is there as an integral part of the rocks.

Of recent years, more and more emphasis has been placed on the water in the rocks. The stimulus has come mainly from petroleum geologists and geophysicists. This is also natural, not only because their prime interest is in the fluids but also because the petroleum geologist these days rarely examines rocks either in hand specimen or in outcrop because the rocks he studies are out of reach at considerable depths below the surface, commonly under the sea. They must rely on indirect data on the rocks *in situ*, mainly in the form of electrical and acoustical measurements. These measurements depend largely on the nature of the fluids in the rocks.

Progress in the understanding of the role of water in geology has been hampered by widespread misunderstanding of the physics involved. Our purpose is to seek a basic understanding of the physics of pore water and its movement, at least to the point where we can use that understanding intelligently in geology and the applications to ground water. In seeking this goal, we acquire the tools we need for an understanding of the more complex role that pore water plays in structural geology – from the deformation of sedimentary basins to the sliding of large rock masses. We therefore start with the most elementary considerations.

Properties of fluids

A fluid is a substance that yields at once to shear or tangential stress. It differs from a solid in a number of obvious respects, but it is important for the geologist to be more precise than the engineer in these concepts. Engineers sometimes define fluids as material that yields *in time* to the slightest shear stress. This definition is perfectly satisfactory provided the time-scale is short; but it obscures the distinction between fluids and solids for geologists because many geological materials that the engineer regards as solid will yield to slight stress over long periods of time. For example, pitch can be broken with a hammer, but it flows (as in Pitch Lake, Trinidad) if smaller stresses are applied for a longer time.

However, we cannot at the outset become distracted from our main purpose, which is to examine the role of fluids, as distinct from solids, in geology. So we shall regard as solid those materials that resist shear stress applied for a short time, measured in hours or days: and we shall regard as fluid those that do not.

This fundamental property of fluids is a consequence of relatively wide molecular spacing, compared to solids. Fluids may be liquids or gases, and these too differ in the matter of molecular spacing, so that gases yield to shear stress more readily than liquids. The molecular spacing also accounts for the fact that gases are more compressible than liquids, and liquids more compressible than solids. By compressible, we mean that a volume of the material can be reduced by the application of an exterior force.

A fluid, as the name implies, can flow. Some fluids flow more readily than others. *Viscosity* (η, ν) is the property of internal resistance to flow; the greater the resistance to flow, the higher the viscosity. An *ideal fluid* is one in which there is no internal resistance to flow: it has no viscosity. Real fluids have viscosity, and this property is important in determining the rate at which fluid flows.

Consider two adjacent layers within a liquid, or a liquid between two closely spaced plates (Fig. 1-1) of which the top plate moves at constant velocity, V, relative to the lower plate. Fluid adheres to each plate, so that the fluid against the upper plate moves at velocity V, while that against the lower plate does not move. Between

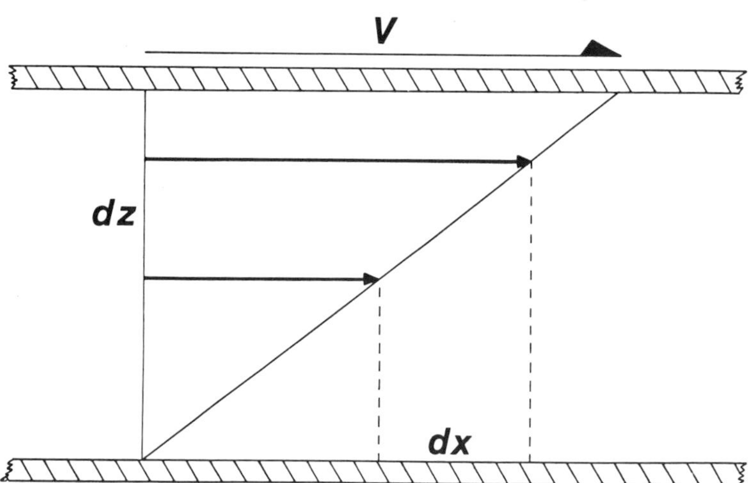

Fig. 1-1. Newtonian concept of viscosity.

the two, we can visualize progressive slippage of one layer over the adjacent layers. This is Newton's conception of viscosity.

The *coefficient of viscosity*, η, (also known as the *absolute* and *dynamic* viscosity) is defined as the ratio of shear stress to the rate of shear strain:

$$\eta = \frac{\tau}{dV/dz}; \qquad \tau = \eta \frac{dV}{dz}. \tag{1.1}$$

In words, the coefficient of viscosity is the tangential force per unit area that maintains unit relative velocity between two parallel planes unit distance apart. It has dimensions $\dfrac{ML^{-1}T^{-2}}{LT^{-1}L^{-1}} = ML^{-1}T^{-1}$.

The unit of viscosity is newton second per square metre ($\mathrm{N\,s\,m^{-2}}$) or the *poise*, P, which is dyne second per square centimetre in units of force, length and time; or grams per centimetre second ($\mathrm{g\,cm^{-1}\,s^{-1}}$) in units of mass, length and time. (The poise is named after the French physician, J.L.M. Poiseuille, 1799-1869, whose interest in blood flow led him to conduct a beautiful series of experiments on flow through small pipes.)

If the coefficient of viscosity remains constant through the fluid, the fluid is said to be *newtonian*, and its viscosity *newtonian*. *Non-newtonian* fluids, of course, are those in which the coefficient of viscosity does not remain constant. We can see the mutual relationships in Fig. 1-2. The ideal fluid is represented by the horizontal line at $\tau = 0$ (no shear stress, finite rate of shear strain). The ideal elastic solid is represented by the vertical line at $dx/dz = 0$ (finite shear stress, zero *rate* of shear strain).

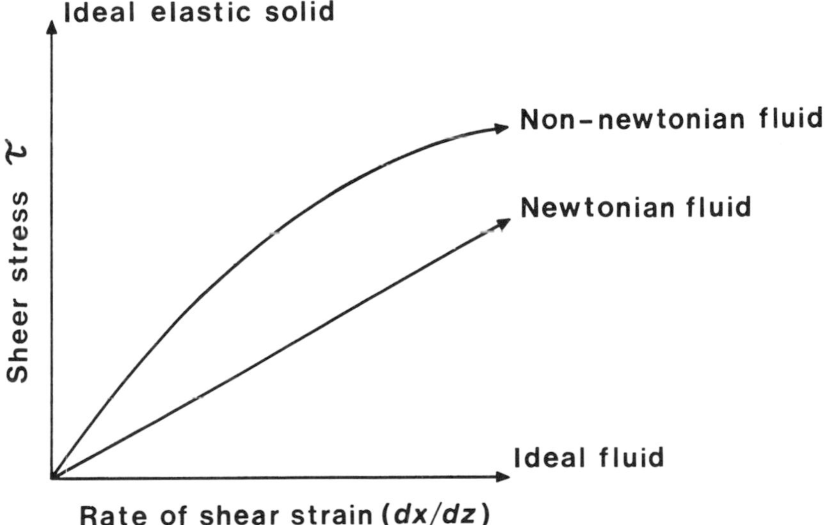

Fig. 1-2. Schematic relationships between shear stress and rate of shear strain for various types of materials.

Newtonian fluids are represented by straight lines passing through the origin, while non-newtonian fluids are represented by curved lines. Viscosity, with time as one of its dimensions, is a concept that links fluids and solids.

The ratio of dynamic viscosity to mass density, η/ρ, appears in many expressions in fluid mechanics. It has dimensions L^2T^{-1}, and is known as *kinematic* viscosity (ν) because it concerns motion without reference to force. The unit of kinematic viscosity is m^2s^{-1} in SI units, or the *stoke*, S, which is cm^2/s. (The stoke is named after the British mathematician and physicist Sir George Stokes, 1819-1903, whose work on the effect of the internal friction of fluids on the motion of pendulums led, amongst other things, to what is now called Stokes' law for the terminal velocity of a sphere falling through a fluid.)

Density (ρ) is defined as the mass of unit volume of the substance. Its dimensions are ML^{-3}. The weight of unit volume of the substance (ρg, where g is the acceleration due to gravity) is called the *specific weight* or *weight density* (γ). Its dimensions are $ML^{-2}T^{-2}$. The unit of mass density (it is better to qualify it so, and avoid ambiguity) is kilogramme per cubic metre (kg m^{-3}) in SI units, or gramme per cubic centimetre (g cm^{-3}). The unit of specific weight or weight density (as distinct from mass density) is newton per cubic metre (N m^{-3}). In many practical applications the specific weight is converted to units of pressure per unit of vertical length, such as kg-f/cm^2 per metre or psi/foot.

The reciprocal of mass density is *specific volume*.

Density also involves a matter of scale in the dimensions of the 'unit volume': it must be large relative to the particle size in order to give a representative figure. For example, consider the density of a heterogeneous material such as a porous sand. The unit volume can be taken so small that it only represents the pore fluid, or the grain material. There is some minimum size that statistically represents the material, with its bulk proportions of solid and fluid.

Specific gravity, the ratio of the density of a substance to that of pure water at a standard temperature (4 °C for scientists, 60 °F for engineers) is a dimensionless quantity that will not be used because it is both inconvenient and commonly imprecise.

The *pressure* (p) at a point in a fluid is defined as the force per unit area on a very small area surrounding the point. Its dimensions are $ML^{-1}T^{-2}$. The SI unit of pressure is the pascal, or newton per square metre (Pa = N m^{-2} = kg $m^{-1}s^{-2}$). Other units in use are kg-f/cm^2 and psi (pounds per square inch).

Stress is also a force per unit area, but it is not strictly synonymous with pressure because the force leading to a state of stress is a surface force that can be resolved into a normal component (σ) and a shear or tangential component (τ). Stress is a tensor quantity, and the definition of a state of stress at a point in a solid requires six quantities to be known (see any text book on structural geology or rock mechanics).

STATICS

The inability of fluids to resist shear stress leads to a number of fundamental propositions.

1) *The free surface of a liquid in static equilibrium is horizontal.* This is a matter of common observation and practical application. To demonstrate this, we postulate an angle of slope θ to the free surface (Fig. 1-3) (the condition of static equilibrium requires the fluid above the free surface to be less dense that the liquid we are considering). The weight W of a small element of the liquid at this surface can be resolved into a normal component, $W \cos \theta$, and a shear component, $W \sin \theta$. Since liquid cannot sustain a shear stress, θ must take the value at which the shear stress is zero, i.e., $\theta = 0$.

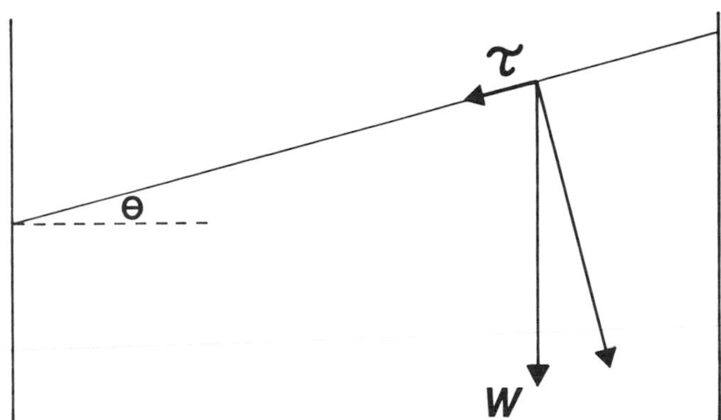

Fig. 1-3.

2) *Surfaces of equal pressure in static fluids are horizontal.* Consider a volume of liquid in a right-cylindrical container (Fig. 1-4a + 1-4b). The weight of this liquid is $\rho g A d$ and it is supported by the base of area A. The pressure on the base is therefore

$$p = \frac{\rho g A d}{A} = \rho g d \tag{1.2}$$

The pressure is dependent only upon the density of the liquid and its depth (we assume g to be constant for practical purposes). If ρ is not constant and not a function of d, shear stresses will exist that will cause flow until they are eliminated and ρ becomes a function of d only. The shape of the container does not alter this (Fig. 1-4b).

3) *The pressure at a point in a liquid in static equilibrium is equal in all directions.* Consider a small prism of the liquid (so small that its weight is insignificant compared to the pressure) of which side b has unit area (Fig. 1-5). The area of side a is then $\sin \alpha$; and of c, $\cos \alpha$. The force (pressure $\times$ area) exerted by the liquid on side a is $p_a \sin \alpha$, and on side c it is $p_c \cos \alpha$. The force p_b on unit side b can be resolved into

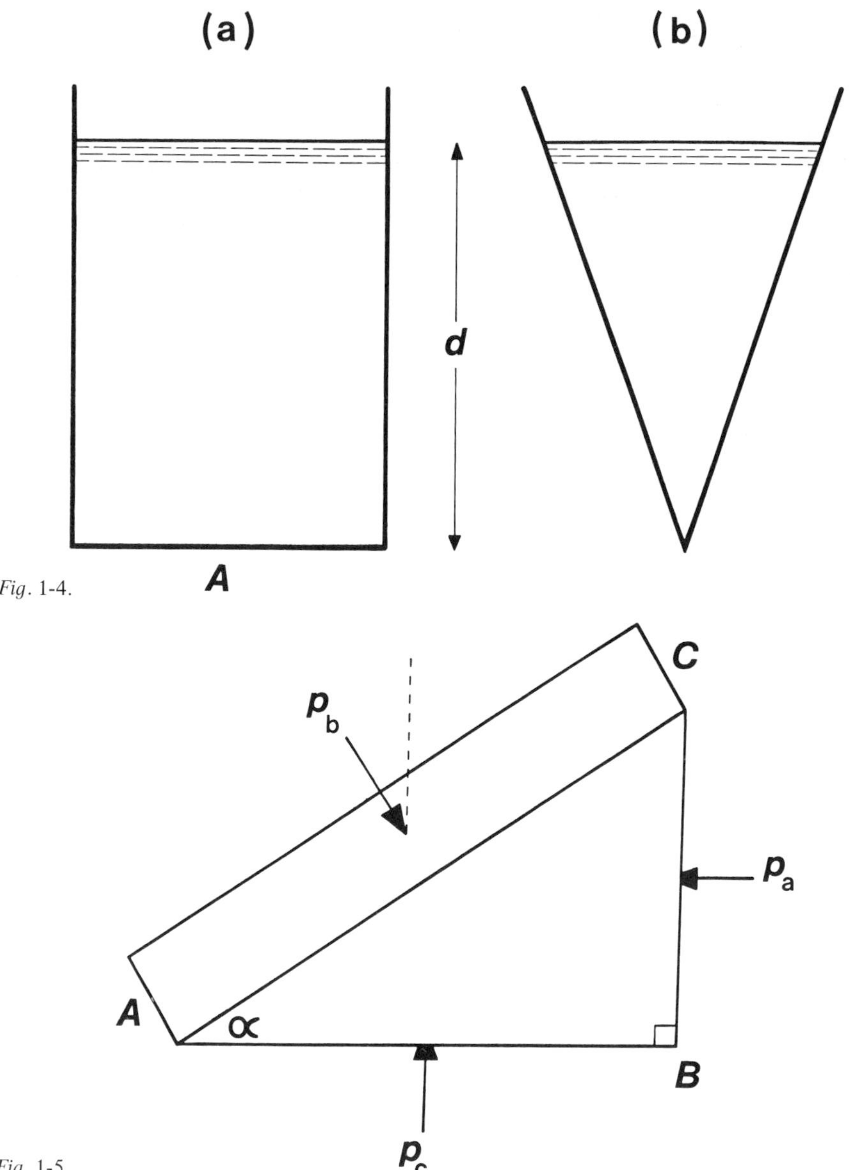

Fig. 1-4.

Fig. 1-5.

horizontal and vertical components, $p_b \sin \alpha$ and $p_b \cos \alpha$ respectively. Since the prism is in static equilibrium, the sum of the forces acting on it must be zero. Hence,

$$p_a \sin \alpha - p_b \sin \alpha = 0; \text{ and } p_a = p_b,$$
$$p_c \cos \alpha - p_b \cos \alpha = 0; \text{ and } p_c = p_b,$$

and the forces acting on the two parallel sides are clearly equal and opposite from the second proposition. Since $p_a = p_b = p_c$ is independent of the angle α, which may arbitrarily assigned a value and orientation, the pressure *at a point* in a static liquid is equal in all directions. (Note carefully that the pressure on a submerged *object* is not equal in all directions.)

This leads to a further proposition:

4) *The horizontal force acting on a surface in a static fluid is the product of the pressure and the vertical projection of the area.* Consider the prism in Figure 1-5 again. The force acting on side b, of unit area, is p_b, the horizontal component of which is as before $p_b \sin \alpha$. But $\sin \alpha$ is the area of side a, which is the area of the projection of b onto a vertical surface. This proposition can also be shown to be true for curved surfaces by considering a very small surface area and its tangential slope.

BUOYANCY

It will be convenient at this point to consider buoyancy, lest the reader think that we are concerned with trivialities. Buoyancy is a superficially simple process that acts on all materials immersed in fluids, and intuitive reasoning can be misleading.

The matter of the pressure exerted by a static liquid on a point P within it is usually demonstrated as follows (Fig. 1-6):

Consider a small vertical cylinder of the liquid of cross-sectional area A, extending from the point P to the surface of the liquid d units above. The forces acting on the vertical surface of the cylinder are all horizontal, and since no shear stress can exist, there is no vertical component. The weight of the cylinder, $\rho g A d$, is therefore supported by an equal force acting upwards on its base,

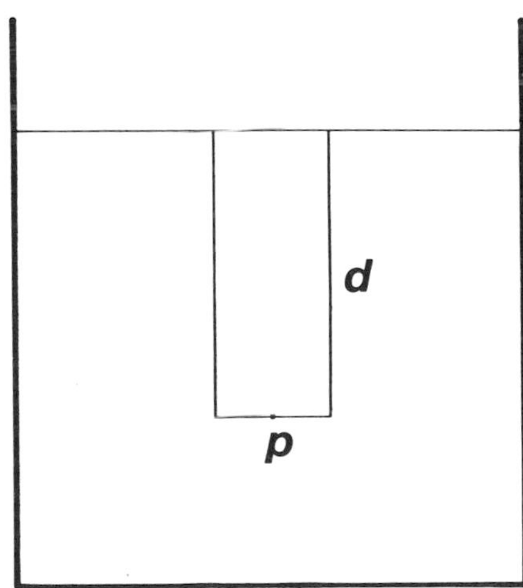

Fig. 1-6.

$$pA = \rho g A d.$$

Hence

$$p = \rho g d,$$

which is the result we had before. Superficially, this seems satisfactory enough. Now extend the argument to the pressure exerted by the liquid at a point on the lower boundary surface, in contact with the container. This pressure (Fig. 1-7) is clearly $\rho g h$; but what do we mean by the force acting on the base of the cylinder when the base of the cylinder is the container?

If a container similar to that in Figure 1-7, but with a very thin, flexible but impermeable base, is inserted into a larger container with identical liquid until the levels in both coincide (Fig. 1-8) the physical situation of Figure 1-6 is restored, and we can visualize the effect in terms of equilibrium and vertically-directed forces. There is really no doubt that the pressure in the liquid at the base of the container is $\rho g h$.

Now, instead of a conceptual cylinder we insert a real solid cylindrical rod of impervious material, with length h and mass density ρ_s that is greater than the density of the liquid, ρ. It is a simple matter to verify that as the rod is inserted, so its weight (as measured by the tension on the top end) decreases. What are the forces acting on the rod when it is in the position comparable to Figure 1-6?

The weight of the rod in air is $\rho_s g A h$. Over the immersed portion, liquid pressure is acting on the vertical surface without shear components, so this cannot reduce the weight. But at the base of the rod there could be an upward vertical force $\rho g A d$. It is a High School experiment to show that the weight on immersion to depth d in the liquid is given by

$$W_d = \rho_s g A h - \rho g A d \tag{1.3}$$

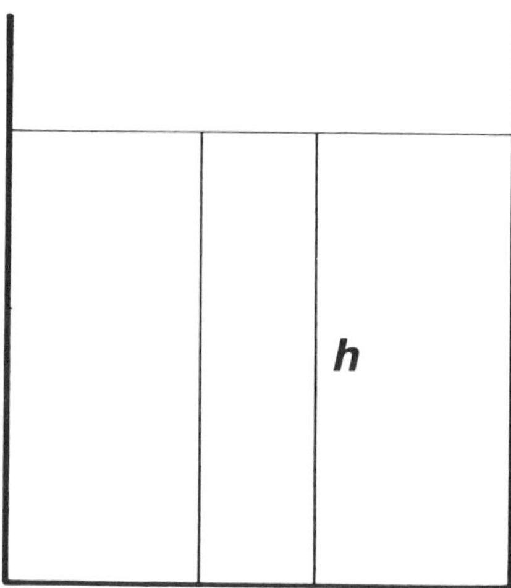

Fig. 1-7.

or, in words, the weight of the rod is reduced by the force due to the liquid at depth d acting upwards on the base of the rod. Alternatively, since ρgAd is the weight of the liquid displaced by the rod, the weight of the rod is reduced by the weight of the liquid displaced (Archimedes' Principle). These are valid alternative *numerical* statements, but are they valid alternative physical statements?

Consider the total immersion of the rod, with the liquid level adjusted to coincide with its top, and the base of the rod now attached to the container in such a way that

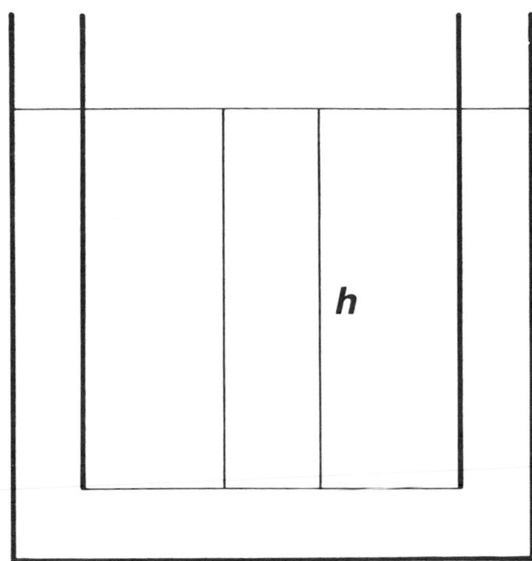

Fig. 1-8.

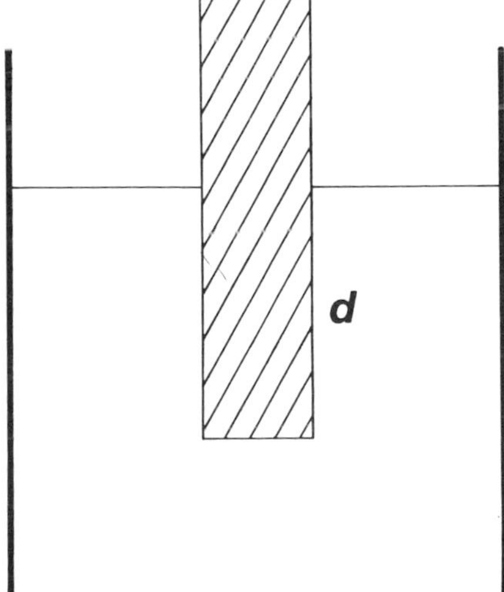

Fig. 1-9. Solid cylinder inserted into liquid.

the liquid is totally excluded, and the rod becomes an integral part of the container (Fig. 1-10). There is no doubt that its weight is now

$$W_h = \rho_s g A h - \rho g A h = (\rho_s - \rho) g A h \tag{1.4}$$

but all forces are apparently normal to the vertical surface of the rod, and there is no basal surface on which the liquid pressure can act: the weight of the rod has been reduced *as if* the liquid acted on the basal surface. The physical validity of the concept of an upward force on the base of the rod is compromised: the real force is due to the density difference, and it acts over the whole submerged volume of the rod. We must regard the hydrostatic pressure field as being continuous throughout the space occupied by the rod, as well as the liquid. It is supporting the partial density ρ of the rod: the remainder, $(\rho_s - \rho)$, is as if it were in a vacuum. Confirmation of this can be found in the observable results of running drillpipe into deep boreholes filled with mud.

In the mid-1950s, development drilling in several oilfields was hampered by what was known as 'wall-sticking' or 'differential sticking'. The pressure of the mud acting on the drillpipe opposite depleted reservoirs with pore pressures considerably below normal hydrostatic, held the pipe firmly to the wall of the hole. To try to free the pipe, it was pulled at the surface. The amount of pull was limited by the tensile strength of the pipe. The question was: what is the weight of the drillpipe above the stuck point? That is, does the force of buoyancy act on the drillpipe above the stuck point? If so, the pipe could be pulled that much harder: if not, as was commonly believed in the early 1950s, the pipe could only be pulled to its tensile strength (less some safety factor).

Consider the forces acting on a open-ended pipe hanging freely in a borehole filled with mud – this is the so-called 'drillpipe fallacy' (see Hubbert and Rubey, 1961,

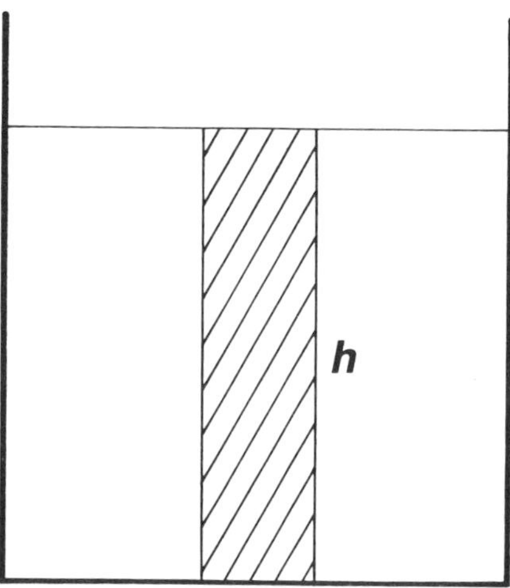

Fig. 1-10. Solid cylinder integral part of container.

p. 1593, for a more detailed discussion). For this, we shall consider 3,000 m of pipe hanging open-ended in mud of mass density 1,500 kg m^{-3} (1.5 g/cm^3). The pipe weighs 26.9 kg/m in air, displaces 3.3×10^{-3} m^3/m, and has a cross-sectional area of 3.3×10^{-3} m^2. The pressure exerted by the mud at the bottom of the pipe is $\rho g d = 1{,}500 \times 9.8 \times 3{,}000 = 4.41 \times 10^7$ Pa (450 kg-f/cm^2). This is providing an upward force of 1.46×10^5 N (14,850 kg-f) on the bottom of the pipe, equivalent to the weight of about 550 m of the pipe in air. Since we can see the bend in 30 m of pipe standing in the rack on the rig floor, it is clear that 550 m of pipe standing on end in air would buckle. But it does not buckle in the borehole; so either the physical interpretation must be wrong or there are factors that have not been taken into account.

Alternatively, the pipe displaces 9.9 m^3 of mud that weights 14,850 kg; so the weight of the pipe in the borehole is 80,700 − 14,850 = 65,850 kg. The *effective* weight per metre is 22.0 kg, and the tensile load on the pipe varies continuously from its total effective weight at the top to zero at the bottom, and the whole length of the pipe is in tension. Buoyancy, like weight, must be a *body* force.

Reverting to the original question, whether the force of buoyancy acts on pipe above the stuck point, the answer is *yes:* the weight of the pipe above the stuck point is reduced by the weight of the mud displaced.

LIQUID-FILLED POROUS SOLIDS IN STATIC EQUILIBRIUM

If we consider unit bulk volume of a porous solid (the volume being large relative to the pore and grain sizes), it will be evident that the proportion of pore space determines the amount of liquid that can be contained. *Porosity* (f) is the ratio of pore volume to bulk or total volume: $v_p/(v_p + v_s)$. In some geological contexts it may be convenient to use the engineers' *void ratio* (ε) which is the ratio of pore volume to the volume of the solids: $f/(1 - f)$. This has the merit of being the ratio of a variable to a constant in most geological contexts.

Porosity has a statistical component grains and pore spaces are never identical, seldom similar. Any arbitrary plane through the material will intersect both grains and pore spaces. In this plane, the area of pore space over the total area is a measure of the porosity of the material. This will vary slightly in planes in different positions and with different orientations.

Porosity and bulk density of a material are related:

$$\rho_b = f\rho_p + (1 - f)\rho_s = \rho_s - f(\rho_s - \rho_g), \tag{1.5}$$

where ρ_p is the pore-fluid density, and ρ_s the grain density. Since the mean grain density is commonly about 2 650 kg m^{-3} (2.65 g/cm^3) an estimate of porosity or bulk density can easily be made.

The propositions treated at the beginning of the chapter have their counterparts in porous solids *provided* the pore spaces are interconnected so that a continuous liquid phase exists. The interconnected porosity is known as *effective* porosity.

Take a container and fill it with loose coarse sand*. Water can be poured into the container to the extent of the pore space before the surface of the water rises above the sand. When the water is at rest, its upper surface is horizontal. This can be measured by inserting manometer tubes (of larger internal diameter than the pore spaces) into the sand at the edge of the container (Fig. 1-11). Since the water cannot sustain a shear stress there will be a uniform pressure field extending throughout the inter-connected pore space, the pressure increasing downwards as the depth. For the same reason, the pressure will be normal to the surfaces of the sand grains.

Close examination of the surface of the grains at the interface between the water and air will, however, reveal that the surface of the water between grains is not horizontal, but concave upwards into contact with the solid particles. This is due to *surface tension*, and is a departure from the fundamental property of liquids that we started with – that the surface of liquid at rest is horizontal. But is is a departure on a small scale that only applies at contacts between solids, liquid and air, or with solids and two different liquids (when it is known as *interfacial tension*). The reality of surface tension can be seen by inserting a tube of very small internal diameter into water, and the water level in this *capillary* tube rises above the surface of the free water (hence the qualification above, that the manometer tube should be of larger diameter than the pore spaces).

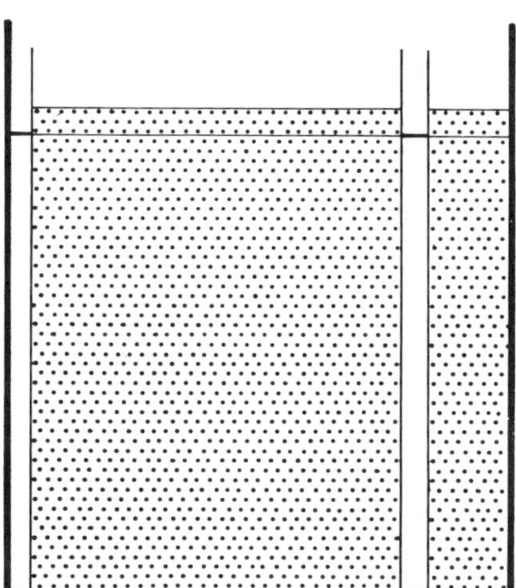

Fig. 1-11. Water-level in porous sand.

* This discussion and these diagrams represent mental experiments, or experiments on a very large scale. Isaac Roberts discovered nearly a century ago that pressure on the bottom of a bin did not increase after the grain in it had reached a depth of more than twice the width of the bin. See Cowin, 1977, for the modern theory.

Surface tension distorts the free surface of water in fine-grained sediments: the finer the grain, the smaller the pore spaces and the greater the elevation of the apparent water level above the level it would have had without surface tension. There are real difficulties in visualizing this effect because, if we regard water as incompressible over the small pressure range involved, and as having negligible tensile strength, there cannot be an increase in the volume of water. The apparent increase must be due to the upper fluid's presence, leading eventually to a transition from 100% water to 100% air over a vertical distance that is determined by the capillary pressure of the water in the material.

Interfacial tension between water and oil or gas is of considerable importance in petroleum reservoir engineering and the theoretical approach to the problems of petroleum migration. However, we shall ignore the effects of surface tension and capillary pressure on the interface between water and air for the time being.

Bulk density

Take a container of known volume. Fill it with a known weight of clean sand, agitating it to achieve a stable packing. Pour in a known weight of water until the water surface coincides with the top of the sand. Clearly the weight of the water-saturated sand exerts a force on the bottom of the container, and that the total force is the sum of the force exerted by the water and that exerted by the sand. (Sand could sustain a slight shear stress, but we shall assume that it does not, and that the whole weight is supported by the bottom of the container.)

The porosity of the sand in the container is the ratio of the pore-water volume to the total volume, v_w/v. The bulk weight is the sum of the weights of solid and water, and can be expressed $\rho_b g v$; and the bulk density is the bulk weight divided by the bulk volume, $\rho_b g$ ($= \gamma_b$). Again, there is a scale effect: the bulk volume considered must be large enough to be representative.

The question arises: what are the relative contributions to the vertical force, and hence pressure, exerted by water-saturated sediment? (To avoid ambiguity, we shall consider the vertical forces on a horizontal plane a few grains above the contact between the base of the sand and the container.)

The total weight of water-saturated sediment in the container is made up of the weight of solids

$$W_s = \rho_s g v (1 - f) \tag{1.6a}$$

and the weight of water in the pores

$$W_w = \rho_w g v f \tag{1.6b}$$

the sum of which is

$$\rho_b g v = W_s + W_w = [\rho_s (1 - f) + \rho_w f] g v. \tag{1.6c}$$

This can be verified experimentally. What about buoyancy? The effective weight of the solids is reduced by the weight of the water displaced. Should we not write for the weight of solids

$$W_s = \rho_s gv(1-f) - \rho_w gv(1-f)$$
$$= (\rho_s - \rho_w)gv(1-f)?$$

Adding this term to $\rho_w gvf$ gives a total that is less than the total weight of the materials put into the container by the amount $\rho_w gv\,(1-f)$, which is the weight of water displaced by the solids. If we had filled the container to the brim with water and then put in the sand, these last expressions would have correctly led to the total weight (this is how the volume of solids is usually measured). By putting the sand into the container first and then adding sufficient water to fill the pore space, the volume of water displaced by the solids is not, in effect, added. Buoyancy does play a part, as we shall see.

Pressure and stress

We have found that the total weight of a water-saturated porous solid is the sum of the weight of the solids and the weight of the pore water, as one would suppose. The question now arises: what are the pressure or stress relationships?

If we divide equation (1.6c) by A, the total area of the base, we get

$$\rho_b gh = \rho_s gh(1-f) + \rho_w ghf \cdot \tag{1.7}$$

This suggests that the pressure or stress due to the solids is only acting on the proportion of A that is occupied by solids, $(1-f)$, and that the water pressure is only acting on the proportion of A occupied by pore space.

Terzaghi (1933) examined the effect of buoyancy on concrete dams. He postulated that buoyancy only acts on solid surfaces exposed to water in the pore spaces, and set up a buoyancy coefficient

$$m = 1 - \frac{A_c}{A},$$

where A_c is the area of solids in the undulating surface S' that only cuts the contacts between grains (Fig. 1-12), projected onto the horizontal plane surface S of area A. Thus the term A_c/A is the proportion of the area not exposed to buoyant forces, and m is a measure of the area over which buoyant forces do act. In two series of experiments he found that the value of m varied but little from unity. Even with dense concrete of low porosity, the buoyant force was found to be

$$F_B = \gamma v(m-f) \simeq \gamma v(1-f),$$

where γ is the specific weight of the water.

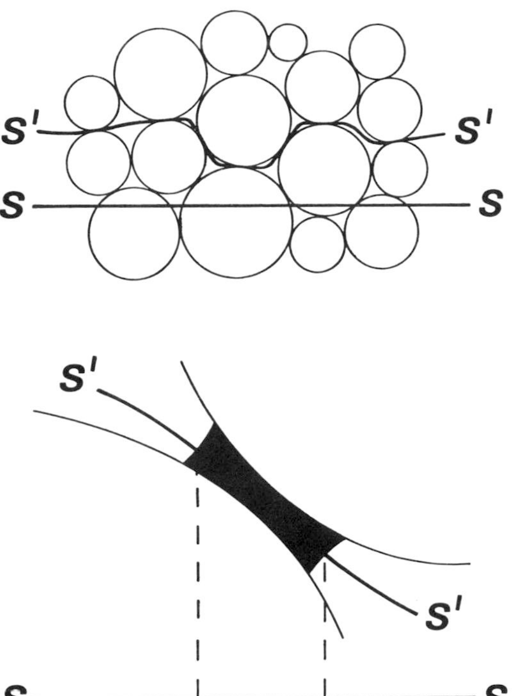

Fig. 1-12. Terzaghi's concept of areas on which pore pressure might be exerted. (After Terzaghi, 1933, fig. 1).

This paradoxical result is of some interest because it means that the force of buoyancy acts on the total area, $(1 - f)$, of solids irrespective of the contact areas between grains. Clearly the value of A_c is not zero because concrete's tensile strength depends on bonding between the grains.

Later, he found the same results in plastic clays (Terzaghi, 1936), and came to the conclusion "The strain in clay and in concrete exclusively depends on the differences between the total stresses and the neutral stresses. In every point of the saturated material the neutral stresses act in every direction with equal intensity and they are equal to the pressure in the water at that point." (Terzaghi, 1936, p. 875). It is hard to escape the conclusion that Terzaghi regarded buoyancy as a body force, and that the field of fluid pressures (neutral stresses) extends throughout the porous solid, through solids as well as the pore liquids*.

* Terzaghi seems to have retreated from this position in later work. Harza's paper on the significance of pore pressure in hydraulic structures, in which he came to conclusions similar to those of Terzaghi (1936), was first submitted for publication in 1937, and resubmitted with revisions 10 years later (Harza, 1949). The discussion takes up more than three times the space of the paper itself, and about half of this is adverse to Harza's thesis. See also Hubbert and Rubey, 1959, pp. 129-142; Hubbert and Rubey, 1960, pp. 621-628; Hubbert and Rubey, 1961, for more detailed and extensive discussion. The view that buoyancy is a surface force and that the fluid pressure acts only on the area fA still appears to be widely held.

The partition of stresses and pressures in porous solids is now generally called *Terzaghi's relationship:*

$$\sigma = S - p, \tag{1.8}$$

where σ is the *effective* stress, S is the total vertical stress ($\rho_b gh$) and p is the fluid pressure or *neutral* stress ($\rho_w gh$ when the pressures are normal hydrostatic).

Reverting now to equation (1.7), we see that this is but one possible partition. Substituting into equation (1.8) the expressions for S and p, we obtain

$$\begin{aligned}\sigma &= \rho_b gh - \rho_w gh = [\rho_s(1-f) + \rho_w f - \rho_w]gh \\ &= [\rho_s(1-f) - \rho_w(1-f)]gh.\end{aligned}$$

The effective stress in porous solids is due to the partial density of solids less the water they displace; and it is this partition of total stress that agrees with experimental results (see also Hubbert and Rubey, 1959, pp. 139-142).

The effective stress, then, is the difference between the total stress and the pore-liquid pressure; and it is important in geology because it is this stress that causes mechanical compaction of sediments under the force of gravity. (The neutral stress is so called because it does not cause deformation.) To explore this further, we shall consider the effective stress in a water-saturated sediment when there is a depth of water on top of the sediment (Fig. 1-13).

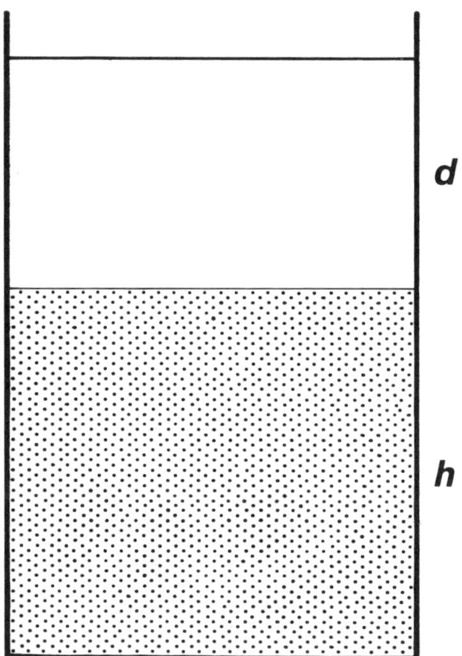

Fig. 1-13.

If the water-saturated sand that we have been considering is overlain by a depth d of water, the total weight W now consists of the weight of water

$$W_w = \rho_w ghAf + \rho_w gAd \tag{1.9a}$$

and the weight of solids

$$W_s = \rho_s ghA\,(1 - f)\,. \tag{1.9b}$$

Adding these, and using equation (1.6c),

$$\begin{aligned} W &= \rho_s ghA\,(1 - f) + \rho_w ghAf + \rho_w gAd \\ &= \rho_b ghA + \rho_w gAd. \end{aligned} \tag{1.9c}$$

The total pressure or stress at the base of the container is $\rho_b gh + \rho_w gd$, and the water pressure is, of course, $\rho_w g(h + d)$; hence the effective stress is

$$\sigma = \rho_b gh + \rho_w gd - \rho_w g\,(h + d) = (\rho_b - \rho_w)gh, \tag{1.10}$$

which is the same as before. In words, the effective stress is independent of the depth of water *over* the porous water-saturated sediment. The pore pressure is higher, the total pressure is higher, but the effective stress remains the same.

This result is important in a number of geological contexts; most obviously, it means that deep-water sediment that is porous and permeable compacts under the force of gravity in the same way and to the same extent as an identical sediment in shallow water.

Finally, consider now the consequences of loading a water-saturated porous sediment, as in Figure 1-14. A rigid, impermeable piston is placed on top of the sediment, and it is loaded. Assuming the container to be rigid, what are the conse-

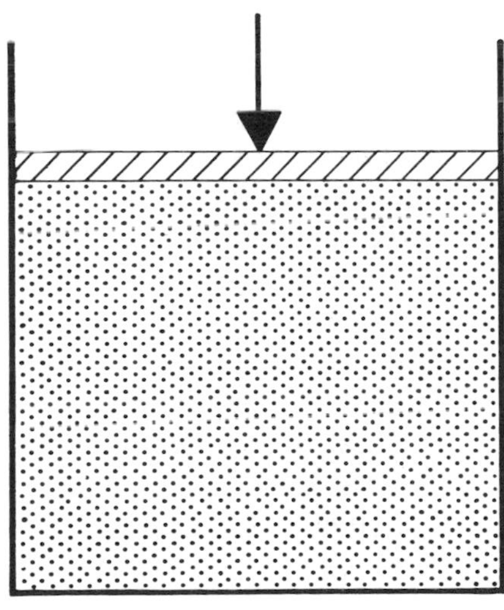

Fig. 1-14.

quences of this loading on the fluids and the solids in the sediment. The answer depends on the compressibility of the solid framework. If the framework is incompressible, it will carry the full load and the pore-fluid pressure remains unaltered. For any degree of compressibility of the solid framework, the consequence of the load F_z will be a rise in the pore-fluid pressure above that given by $\rho_w gh$: the pore fluid will bear part of the load. The least compressible component bears most of the load.

Field confirmation of Terzaghi's relationship

Terzaghi's partition of total stress in a rock into effective stress (σ) and neutral stress or pore pressure (p) is not merely a theoretical conception: evidence in its support comes not only from the experiments of Terzaghi and others, but also from field observations in context of ground water and petroleum.

Before the turn of the century, it was discovered that the load of a train on the railway line near a well could raise the level of the water in the well (King, 1892, p. 67; Veatch, 1906). More recently, Jacob (1939) reported on this effect near the Smithtown railway station, Long Island, New York. The well is located 16.5 m (54 ft) north of the main line, 110 m (360 ft) east of the station. It is 27.1 m (89 ft) deep, with the water-level 10.7–12.2 m (35-40 ft) below the surface. The sequence of sediments in a nearby well is sand (12.2 m) on clay (9.1 m) on gravel (at least 9 m), which is the aquifer. The aquifer is seen to be confined.

Each day there were 11 eastbound and 11 westbound trains weighing anything from 260,000 to 540,000 kg. The same train returns, so the average weight in each direction was about equal. It was found that the westbound trains caused the water-level to rise 9.1 mm, on average, while the eastbound trains caused it to rise 13.7 mm. The difference between the two seems to be due to the different speeds – westbound trains averaging about 12–15 m s^{-1} past the well; eastbound, about 6 m s^{-1}.

Of particular interest are the records of the arrival and departure of a freight train with 19 wagons and a caboose. The train stopped, the locomotive and four wagons were detached, one wagon was shunted to a siding, the locomotive and four wagons were reconnected, and the train left. The changes in water-level during these operations were recorded, and are shown in Figure 1-15.

It is evident that the weight of the train added to that of the overburden increased the effective stress and so reduced the pore volume. The symmetry of the changes associated with arrival and departure suggest elastic recovery of the aquifer.

On a larger scale, we find that the abstraction of ground water in the Houston district of Texas (Gabrysch, 1967) between 1890 and 1961 led to a fall in the static water levels. Over the years 1943 to 1964, the land surface subsided up to 1.5 m (5 ft). Figure 1-16 shows maps of the decline in static water level and of the subsidence, and they leave little doubt that the decline of pore-fluid pressure led to an increase in effective stress, further compaction of the aquifer, and surface subsidence.

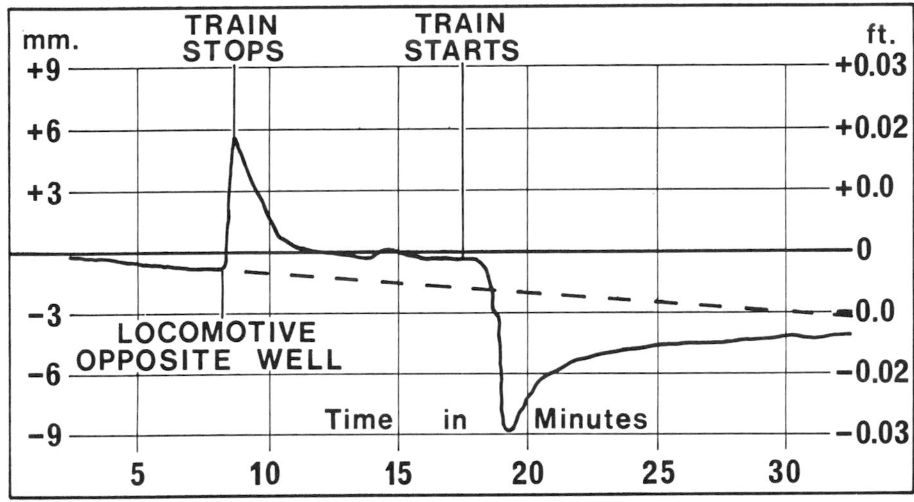

Fig. 1-15. Fluctuations in water-level in a well due to a train (after Jacob, 1939, fig. 5).

The Wilmington oil field in California, discovered in 1932, is one of the world's giant oil fields. It lies below the southern coastal districts of Los Angeles and the city of Long Beach. This is an industrial (and residential) area that includes Long Beach harbour and naval shipyard. During the first 30 years to the end of 1967, more than 1.84×10^8 m^3 (1.156×10^9 bbl) of oil and 23.8×10^9 m^3 (840×10^9 ft^3) of gas were produced (Mayuga, 1970). Its importance from our point of view is that within a decade of major production (which began about 1937, when the significance of the discovery became apparent) subsidence of the land surface began threatening coastal installations. Total subsidence up to 1967 (including a little due to ground-water extraction) is shown in Figure 1-17, and the deepest part of the depression so formed lies over the crest of the oil field. Maximum subsidence amounted to 8.8 m (29 ft). This vertical movement also led to horizontal movements up to 3 m, which also caused extensive damage on the surface and to oil wells at depth. Measurements in boreholes (see Mayuga, 1970, for details) indicate that almost all the compaction had taken place in the producing zones.

The correspondence between production rate and subsidence rate left no doubt as to the cause of the subsidence, for the maximum subsidence rate of 0.71 m/year (2.3 ft/year) occurred in 1951 about 9 months after the production rate peak.

Pilot water-injection schemes, intended both to halt subsidence and to improve oil production, began in 1953; and in 1958, when the major scheme started, production could be increased while the rate of subsidence further decreased. In areas of greatest water injection, the subsidence was even reversed, with raising of the surface

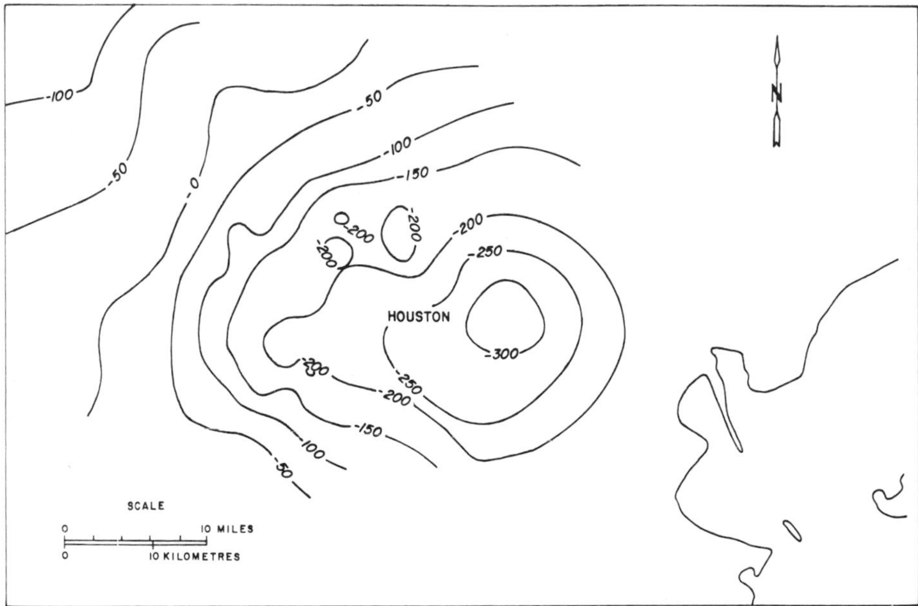

Fig. 1-16a. Decline in static water-level (in feet) in the Houston district, 1943-1964 (after Gabrysch, 1967, fig. 4, by courtesy of the Texas Department of Water Resources).

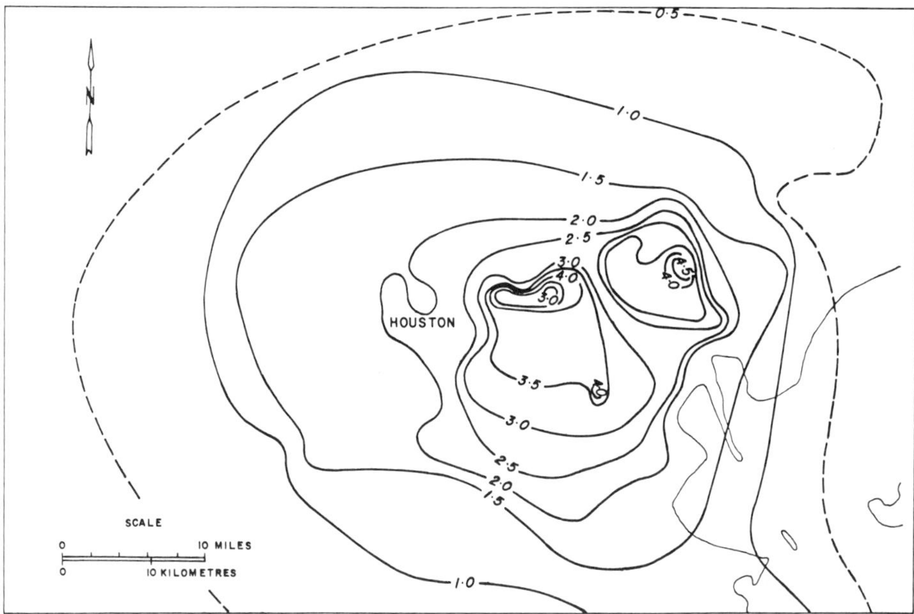

Fig. 1-16*b*. Subsidence of land surface (in feet) in same area as Figure 1-16a over same period (after Gabrysch, 1967, fig. 19, by courtesy of the Texas Department of Water Resources).

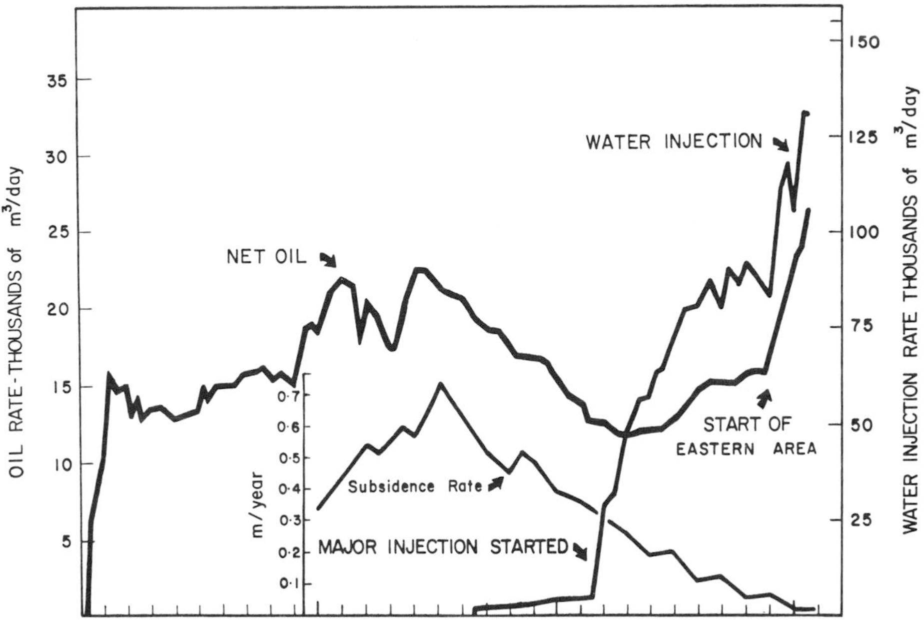

Fig. 1-17. Wilmington oil field, California: oil production, water injection and land subsidence (after Mayuga, 1970, fig. 16).

by up to 25 cm*. A full account of this field is due to Mayuga (1970).

These three examples show that loading a porous and permeable sedimentary rock reduces the pore volume, that abstraction of liquids leads to compaction, and that injection of liquids can de-compact a sedimentary rock. There is little doubt that Terzaghi's relationship is real (at least qualitatively) and that it is the effective stress that compacts detrital sedimentary rocks mechanically.

Quantifying this aspect of geology is not easy. The deformation of solids can be characterized by parameters known as the *elastic moduli* (see any modern text-book on structural geology), which can be determined by laboratory measurement. But the dimension of time is critically important. Elastic behaviour on a short time-scale may well be replaced by plastic or fluid behaviour on a longer time-scale. With porous materials, this means that as the period of time under load increases, so the compressibility of the solid framework increases (the compressibility of fluids does not change with time).

* It is normal practice now to maintain oil reservoir pressures by water injection, both to prevent subsidence and to maintain production. An earlier example of oilfield subsidence is mentioned by M. ap Rhys Price in his discussion of a paper by Kugler (1933, p. 769). Subsidence over the Lagunillas field on the shore of Lake Maracaibo, Venezuela, was found to be 'in direct proportion with the production taken out'.

DIMENSIONAL ANALYSIS

We introduce this powerful tool of fluid mechanics now because readers may be unfamiliar with it, and an elementary understanding of it helps in the handling of the various equations we shall be using.

If an equation is to be generally valid, we must be able to use any consistent set of units, and any constants must be dimensionless. Such an equation is said to be dimensionally homogeneous. All physical quantities, such as volume, density, velocity, can be expressed in terms of the basic units of *M*ass, *L*ength, and *T*ime. Thus, velocity is distance divided by time, or in the notation for dimensions, LT^{-1}. Any equation relating such quantities must be such that exponents or powers of M, L, and T are equal on both sides of the equation. For example,

weight density	=	*mass density*	×	*acceleration of gravity*
γ	=	ρ	×	g

or, in dimensions,

$$ML^{-2}T^{-2} \quad = \quad ML^{-3} \quad \times \quad LT^{-2}.$$

To check this, we set up what are called *indicial equations.*

For M: $1 = 1$

For L: $-2 = -3 + 1$

For T: $-2 = -2.$

The equation thus checks, and it is therefore dimensionally homogeneous. Dimensional analysis uses the requirement of homogeneity to predict the form of an equation relating various physical quantities. An example will make this clear.

Imagine we did not know the relationship between pressure and depth in an incompressible fluid. We might well assume that the pressure p is a function of mass density ρ, g the acceleration due to gravity, and the volume, v (Fig. 1-18), and write

$$p = \psi(\rho, g, v).$$

This equation can be written in dimensional form:

$$\begin{aligned} ML^{-1}T^{-2} &= (ML^{-3})^{a} \quad (LT^{-2})^{b} \quad (L^{3})^{c} \\ &= M^{a}L^{-3a} \quad L^{b}T^{-2b} \quad L^{3c}, \end{aligned}$$

from which we write the indicial equations

for M: $1 = a$

for L: $-1 = -3a + b + 3c$

for T: $-2 = -2b.$

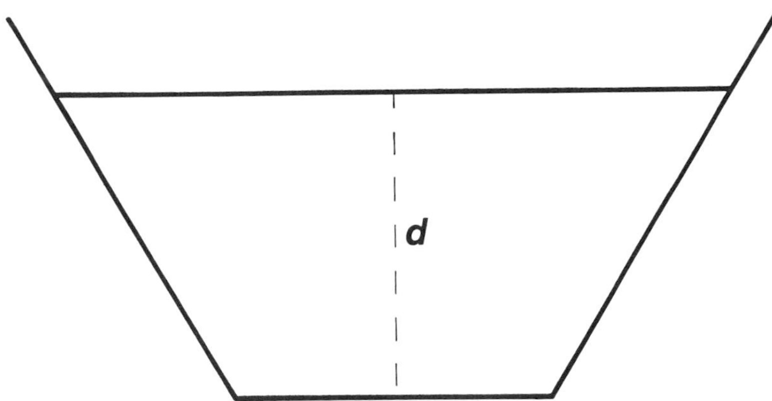

Fig. 1-18.

The solution of these simultaneous equations is, a = 1, b = 1, and the substitution of these in the indicial equation for L gives c = $\frac{1}{3}$. Hence the equation in dimensional form is

$$ML^{-1}T^{-2} = (ML^{-3})\,(LT^{-2})\,(L).$$

The (L) is clearly vertical depth, so the functional relationship between pressure, density, acceleration due to gravity, and depth is

$$p = \text{B}\rho g d\,,$$

where B is a dimensionless constant. A few experimental measurements would establish that the dimensionless quantity

$$p/\rho g d - \text{B} = 1$$

for a liquid of constant density.

All valid equations can be written in dimensionless terms, with the great advantage in experimental work that only two variables in each have to be determined for different values, and the simplest can be chosen.

While dimensional analysis does not necessarily give the full function, it indicates the form it will take. It will have been noted that our original assumption that volume would be a parameter above was rejected by the analysis.

A list of common physical quantities and their dimensions are given in Table 1-1. The reader is referred to Buckingham (1914, 1921), Brinkworth (1968, pp. 79-89), or Pankhurst (1964) for a more detailed account of dimensional analysis.

Table 1.1. Dimensions of some common physical quantities.

Quantity	Dimensions
Acceleration	$L\ T^{-2}$
Bulk modulus (& other moduli)	$M\ L^{-1}T^{-2}$
Compressibility	$M^{-1}L\ T^{2}$
Density, mass	$M\ L^{-3}$
Density, weight	$M\ L^{-2}T^{-2}$
Energy	$M\ L^{2}\ T^{-2}$
Force	$M\ L\ T^{-2}$
Permeability, coefficient of / Hydraulic conductivity	$L\ T^{-1}$
Permeability, intrinsic	L^{2}
Pressure, stress	$M\ L^{-1}T^{-2}$
Specific discharge	$L\ T^{-1}$
Surface tension	$M\ T^{-2}$
Velocity	$L\ T^{-1}$
Viscosity, absolute or dynamic	$M\ L^{-1}T^{-1}$
Viscosity, kinematic	$L^{2}\ T^{-1}$
Weight	$M\ L\ T^{-2}$
Work	$M\ L^{2}\ T^{-2}$
Temperature	$L^{2}\ T^{-2}$
Quantity of heat	$M\ L^{2}\ T^{-2}$
Thermal conductivity (thermal units)	$M\ L^{-1}T^{-1}$

GENERAL REFERENCES

Chow, V.T.,1959. *Open-channel hydraulics*. McGraw-Hill, New York, 680 pp.

Daugherty, R.L., and Franzini, J.B., 1965. *Fluid mechanics with engineering applications* (6th edition). McGraw-Hill, New York, 574 pp. (International Student Edition.)

Forchheimer, P., 1914. *Hydraulik*. B.G. Teubner, Leipzig and Berlin, 566 pp.

Franks, F., (Ed.), 1972. *Water: a comprehensive treatise, Vol. 1. The physics and physical chemistry of water*. Plenum Press, New York and London.

Giles, R.V., 1962. *Theory and problems of fluid mechanics and hydraulics* (2nd edition). McGraw-Hill, New York, 274 pp. (Schaum's Outline Series.)

Riddick, J.A., and Bunger, W.B., 1970. *Organic solvents* (3rd edition). *In:* A. Weissberger (Ed.), *Techniques of chemistry, Vol.* 2. Wiley-Interscience, New York, London, etc.

Robinson, J.L., 1963. *Basic fluid mechanics*. McGraw-Hill, New York, Toronto, and London, 188 pp.

Sorby, H.C., 1908. On the application of quantitative methods to the study of the structure and history of rocks. *Quarterly Journal Geol. Soc. London*, 64: 171-233.

Terzaghi, K., 1943. *Theoretical soil mechanics*. John Wiley & Sons, New York; Chapman & Hall, London, 510 pp.

Terzaghi, K., and Peck, R.B., 1948. *Soil mechanics in engineering practice*. John Wiley & Sons, New York; Chapman & Hall, London, 566 pp.

Tuma, J.J., 1976. *Handbook of physical calculations*. McGraw-Hill, New York, 370 pp.

SELECTED BIBLIOGRAPHY

Bridgman, P.W., 1926. The effect of pressure on the viscosity of forty-three pure liquids. *Proc. American Academy Arts Sciences*, 61(3): 57-99.

Brinkworth, B.J., 1968. *An introduction to experimentation*. English Universities Press, London, 182 pp.

Buckingham, E., 1914. On physically similar systems; illustrations of the use of dimensional equations. *Physical Review*, 2nd series, 4: 345-376.

Buckingham, E., 1921. Notes on the method of dimensions. *Lond. Edinb. Dubl. Phil. Mag.*, 6th series, 42: 696-719.

Cowin, S.C., 1977. The theory of static loads in bins. *J. Applied Mechanics*, 44: 409-412.

Gabrysch, R.K., 1967. Development of ground water in the Houston District, Texas, 1961-65. *Texas Water Development Board, Report* 63: 35 pp.

Harza, L.F., 1949. The significance of pore pressure in hydraulic structures. *Trans. American Soc. Civ. Engineers*, 114: 193-214. (Discussion: 215-289.)

Hubbert, M.K., and Rubey, W.W., 1959. Role of fluid pressure in mechanics of overthrust faulting, I. Mechanics of fluid-filled porous solids and its application to overthrust faulting. *Bull. Geol. Soc. America*, 70 (2): 115-166.

Hubbert, M.K., and Rubey, W.W., 1960. Role of fluid pressure in mechanics of overthrust faulting: a reply [to Laubscher, 1960]. *Bull. Geol. Soc. America*, 71 (5): 617-628.

Hubbert, M.K., and Rubey, W.W., 1961. Role of fluid pressure in mechanics of overthrust faulting: a reply to discussion by Walter L. Moore. *Bull. Geol. Soc. America*, 72 (10): 1587-1594.

Jacob, C.E., 1939. Fluctuations in artesian pressure produced by passing railroad-trains as shown in a well on Long Island, New York. *Trans. American Geophysical Union*, 20: 666-674.

King, F.H., 1892. Observation and experiments on the fluctuations in the level and rate of movement of ground-water on the Wisconsin Agricultural Experiment Station Farm and at Whitewater, Wisconsin. *U.S. Dept. Agriculture, Weather Bureau, Bulletin* no. 5 (75 pp.)

Kugler, H.G., 1933. Contribution to the knowledge of sedimentary volcanism in Trinidad. *J. Instn Petroleum Technologists*, 19: 743-760. (Discussion: 760-772.)

Laubscher, H.P., 1960. Role of fluid pressure in mechanics of overthrust faulting: discussion. *Bull. Geol. Soc. America*, 71 (5): 611-615.

Leliavsky, S., 1947. Experiments on effective uplift area in gravity dams. *Trans. American Soc. Civ. Engineers*, 112: 444-487.

Mayuga, M.N., 1970. Geology and development of California's giant – Wilmington oil field. *In*: M.T. Halbouty (Ed.), Geology of giant petroleum fields. *Memoir American Ass. Petroleum Geologists*, 14: 158-184.

Moore, W.L., 1961. Role of fluid pressure in overthrust faulting: a discussion. *Bull. Geol. Soc. America*, 72 (10): 1581-1586.

Pankhurst, R.C., 1964. *Dimensional analysis and scale factors*. Chapman & Hall, London, 152 pp. (Inst. Physics Monographs for Students.)

Pratt, W.E., 1927. Some questions on the cause of the subsidence of the surface in the Goose Creek field, Texas. *Bull. American Ass. Petroleum Geologists*, 11 (8): 887-889.

Pratt, W.E., and Johnson, D.W., 1926. Local subsidence of the Goose Creek oil field. *J. Geology*, 34 (7): 577-590.

Skempton, A.W., 1970. The consolidation of clays by gravitational compaction. *Quarterly Journal Geol. Soc. London*, 125 (for 1969): 373-411.

Terzaghi, K., 1933. Auftrieb und Kapillardruck an betonierten Talsperren. 1er *Congrès des Grands Barrages* (Stockholm, 1933), 5: 5-15.

Terzaghi, K., 1936. Simple tests determine hydrostatic uplift. *Engineering News Record*, 116 (June 18): 872-875.

Veatch, A.C., 1906. Fluctuations of the water level in wells, with special reference to Long Island, New York. *U.S. Geol. Surv. Water Supply and Irrigation Paper* 155 (83 pp.).

2. LIQUIDS IN MOTION

Just as it is our experience that the upper surface of a liquid at rest is horizontal, so is it our experience that the upper surface of a liquid in motion is inclined from the horizontal, and inclined in the direction of flow. Hydrostatics, the science of fluids at rest, is a special case of hydrodynamics, the science of fluids in motion.

The flow of water in an open channel is governed by two principal forces: the component of the force of gravity along the channel bed tending to accelerate the water, and the frictional resistance between the water and the material of the channel tending to slow it down. The water accelerates until these two forces are equal, and it then flows at a constant rate (in terms of the quantity of water passing a transverse plane in unit time).

All liquid flow falls into one of two categories: laminar flow, when the paths followed by the particles are essentially parallel, and turbulent flow. In geology we are more concerned with laminar flow, although turbulent flow is important in sediment transport mechanisms. The criterion that determines whether flow is laminar or turbulent is a function of the velocity* of the water and the dimensions of the channel. For the present we shall only consider laminar flow, and leave discussion of this criterion for pipe flow.

Water overflowing a dam into an open channel (Fig. 2-1) will accelerate until the resisting forces become equal to the propelling forces, and some time and distance is required for this to be achieved. Flow nearest the dam is *nonuniform;* the water is accelerating and its depth in the channel is decreasing in the direction of flow. As soon as the resisting forces become equal to the propelling forces,the flow becomes *uniform*, with constant rate or velocity, and constant depth. Uniform flow persists until the conditions change – such as a change in slope or of channel dimensions – but will be re-established in these new conditions if they persist for long enough. Flow may also be described as *steady* if, in any position, the velocity remains constant with time.

The study of liquids in motion centres around three basic variables: a characteristic dimension of the flow, the velocity or rate of flow (which also involve the physical properties of the liquid, mainly viscosity), and the hydraulic gradient.

* A useful distinction between velocity and speed is not always easily maintained, as we shall see. Here, the velocity is the volumetric flow rate divided by the transverse area.

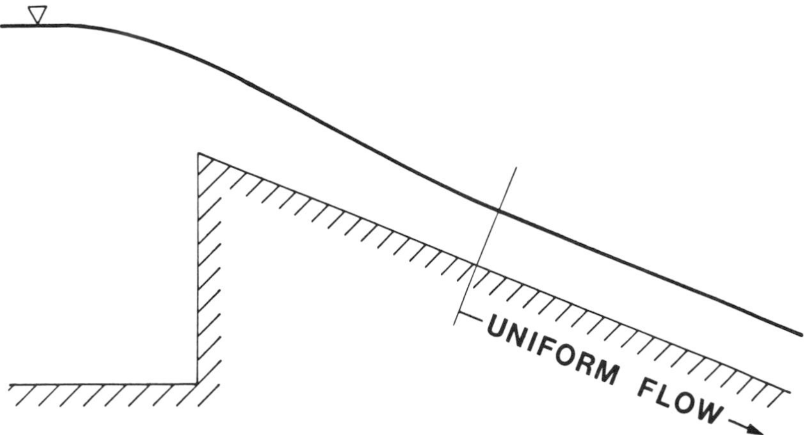

Fig. 2-1. Schematic section through water overflowing a dam into an open channel.

Analysis of liquid flow in pipes and channels proceeds on the assumption that where the liquid is in contact with a solid, the relative velocity of the liquid is zero (as we saw in the discussion of viscosity on p. 2. Consider liquid flow down a straight channel, as in Figure 2-2. The motive force is the component of weight of the water along thc channcl, and it is resisted by the shear forces generated from the confining solid surfaces. When these two forces are equal and opposite, the water ceases to accelerate and the flow is said to be uniform. Then,

$$lwh\,\rho g \sin\theta - \tau_0\, l(w + 2h) = 0, \tag{2.1}$$

where τ_0 is the boundary shear stress and w and h are the width and depth of water (normal to the surface) in a reach of the channel of length l. The characteristic dimension of the channel has been found to be the ratio of boundary shear stress to the component of unit weight down the channel,

$$\begin{aligned}\tau_0/\rho g \sin\theta &= lwh/l(w + 2h)\\ &= \text{volume/wetted surface area}\\ &= \text{area of flow/wetted perimeter}.\end{aligned} \tag{2.2}$$

This is known as the *hydraulic radius* (sometimes, the hydraulic mean depth because this quantity approaches the depth in a wide, shallow channel), and is given the symbol R. It has the dimension of *length*. It is by mcans of the hydraulic radius that flow of different cross-sectional shapes is compared.

Velocity or rate of flow: the speed of a particle of water or other liquid is a variable of position and depends partly on the viscosity of the liquid. It can usually be seen that the speed of the water at the surface near the sides of a channel is less than that near the middle (Fig. 2-3a). This profile, usually called a velocity profile, results from the

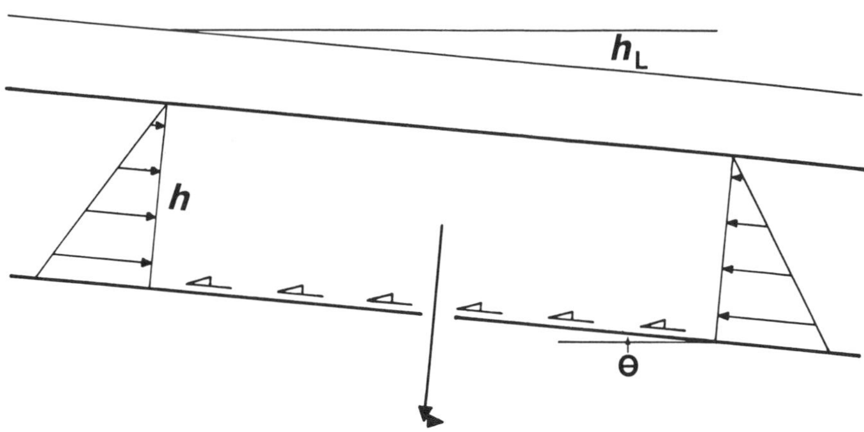

Fig. 2-2. Forces acting on unit length of liquid in uniform laminar flow in an open channel.

frictional resistance along the sides of the channel, and its shape is a function of the viscosity of the liquid, the width of the channel, and its slope. It will readily be appreciated that it is also a function of the depth of the channel, and that there are also analogous profiles in planes normal to the water surface, parallel to the channel sides (Fig. 2-3b). These complexities dictate the practical expedient of defining *mean velocity* as the volumetric flow rate divided by the transverse area of flow. It has the dimensions of velocity ($q = Q/A$: $LT^{-1} = L^3T^{-1}/L^2$). This is satisfactory for steady flow and for all computations involving quantities proportional to velocity; but for quantities such as momentum, which varies as the square of velocity, the velocity profile should be taken into account.

The third basic variable, *hydraulic gradient*, is concerned with the energy gradient down which the liquid flows. Energy exists in various forms, but it is measurable by the amount of work (force × distance moved) that it can do, and it is relative. It has dimensions ML^2T^{-2}. Water has three principal energies: *kinetic*, due to its movement; *potential*, due to its elevation in the gravitational field; and *thermal*, due to its heat. To these we add *pressure* because it can be considered as an energy (work done in compression) and it has the dimensions of energy/volume. Water moves unless it is in mechanica equilibrium with its physical environment and constraints. The principle formulated by Lagrange, late in the 18th Century, is helpful: a mechanical system is in stable equilibrium when the kinetic energy is zero and its potential energy is at a minimum with respect to its physical environment and constraints. For example, if you pour water into a basin, the kinetic energy will be converted into heat by friction and will decrease; mechanical equilibrium will be achieved when the water is at rest and occupying the space of minimum potential energy within the basin. We must clarify the concept of potential energy.

When it has come to rest, the water that was poured into the basin occupies a position that is determined by the shape (i.e., constraints) of the basin, and the

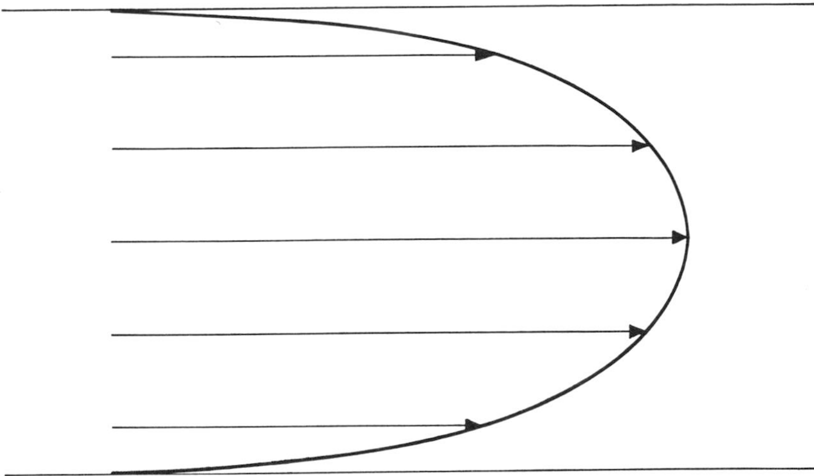

Fig. 2-3a. Speed of water at surface is less at the sides (plan).

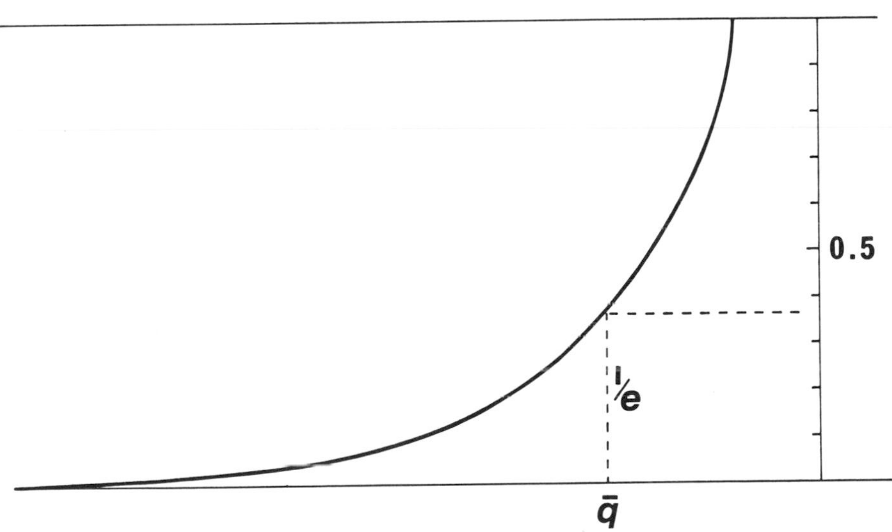

Fig. 2-3b. Speed of water at the bottom is less than that at the top (profile). The mean velocity ($\bar{q}$) is approximately the velocity at $1/e$ of the depth from the bottom.

density of the water relative to the other fluid (air). Although it has come to rest, the water has an energy due to its position in the gravitational field: if the plug is removed, it will 'seek a lower level' and in so doing, acquire kinetic energy. The energy the water has in the basin is not used while the plug remains, but it is potential energy, and it is potential energy due to position in the gravitational field.

Energy, as be have seen, is measurable by the amount of work it can do, and is relative. An identical basin of water at a higher elevation in the earth's gravitational field has a higher potential energy because it can do relatively more work. If it were connected to the lower basin, and the upper plug removed, the water from the upper would displace that of the lower basin, and the combined volume of water would eventually reach mechanical equilibrium (unless it overflowed, in which case some of the potential energy would be realized as kinetic energy). We can therefore take an arbitrary horizontal plane as datum, and the potential (mechanical) energy of unit mass of water would be given by

$$E_p = gz. \qquad (\mathrm{N\,m} = \mathrm{kg\,m^2\,s^{-2}}) \qquad ML^2T^{-2}$$

This is equal to the work required to move unit mass of water a vertical distance z from the reference or datum plane to the position it occupies, by a frictionless process.

If there are changes of pressure and volume, then work is involved in these changes. The movement itself involves kinetic energy; and the work done in overcoming frictional resistance will generate heat (which is a dissipation of mechanical energy). Hence, the water in the basin has potential energy due to its position, and if it is free to flow, this energy will be converted to kinetic energy and heat. This heat loss is irreversible, so that flowing water continuously loses energy.

Bernouilli's theorem

If water is regarded for practical purposes as being incompressible over the pressure changes that may take place in relatively short and slow movement, the principles of conservation and equivalence of energy allow the formulation of an energy balance.

Consider an ideal liquid flowing from station 1 to station 2 along a pipe of decreasing diameter, as in Figure 2-4. The work done on the liquid in the pipe is equal to the change of energy of the liquid. At station 1, the force applied to the liquid is equal to $p_1 A_1$; and in moving the liquid a small distance δl_1 in the small interval of time δt, the work done is equal to $p_1 A_1 \delta l_1$. The weight of liquid passing

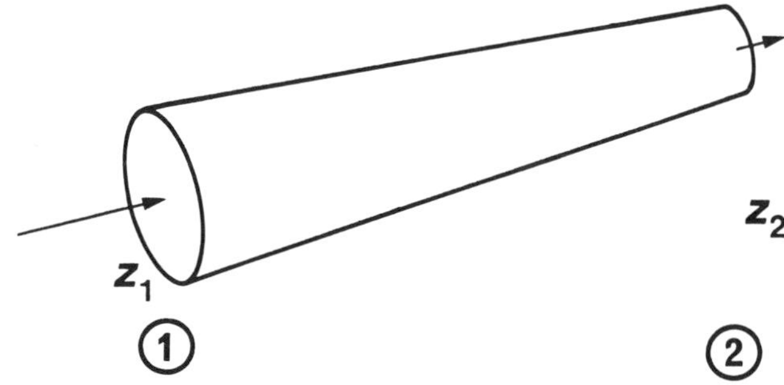

Fig. 2-4. Liquid flowing along converging pipe.

station 1 during the interval of time δt is $\rho g A_1 \delta l_1$; and the same weight of liquid leaves station 2, that is, $\rho g A_2 \delta l_2$.

The potential energy E_p per unit of mass of liquid entering at station 1 is gz_1, so the potential energy of unit weight is z_1 and the potential energy of the liquid entering at station 1 is $\rho g A_1 \delta l_1 z_1$ and that leaving at station 2 is $\rho g A_2 \delta l_2 z_2$.

The kinetic energy E_k per unit of weight is proportional to $q^2/2g$, so the kinetic energy of the liquid entering at station 1 is $\rho A_1 \delta l_1 q_1^2/2$, and that leaving at station 2 is $\rho A_2 \delta l_2 q_2^2/2$.

Finally, there is thermal energy E_t. The effect of friction is to dissipate mechanical energy by raising the temperature of the liquid, with some loss of heat to the pipe – but we are considering an ideal liquid of zero viscosity.

The principle of conservation and equivalence of energy requires that the sum of the work done, the potential energy, and the kinetic energy, shall be constant, i.e.,

$$p_1 A_1 \delta l_1 + \rho g A_1 \delta l_1 z_1 + \rho g A_1 \delta l_1 q_2^2/2g = p_2 A_2 \delta l_2 + \rho g A_2 \delta l_2 z_2 + \rho g A_2 \delta l_2 q_2^2/2g$$

(in which each term has the dimensions of energy, ML^2T^{-2}).
Since $\rho g A_1 \delta l_1 = = \rho g A_2 \delta l_2$,

$$\frac{p_1}{\rho g} + z_1 + \frac{q_1^2}{2g} = \frac{p_2}{\rho g} + z_2 + \frac{q_2^2}{2g}$$

or,

$$\frac{p}{\rho g} + z + \frac{q^2}{2g} = \textit{constant}. \tag{2.3}$$

This is known as Bernouilli's theorem, and it is valid for incompressible, frictionless liquids. Each term has the dimension of *length*. The reader will be uncomfortable over the term $q^2/2g$, wondering what meaning is to be attached to the velocity in view of the definition of mean velocity (p. 28). Strictly, there is a coefficient in this term, to take the velocity profile into account; and the value of this coefficient in laminar pipe flow is 2. But it is normally neglected, and the mean velocity q taken for this term.

The first term in equation (2.3), $p/\rho g$, is known as the *pressure head* and is the height of a column of liquid of density ρ that can be supported by a pressure p. The second term, z, is merely the elevation of the element of liquid above (or below) an arbitrary horizontal datum plane: it is called the *elevation head* or, less desirably, the *potential head*. The third term, $q^2/2g$, is known as the *velocity head*. The sum of these is known as the *total head*; and with an ideal, incompressible liquid of zero viscosity, the total head would remain constant. In ground-water geology, the total head is commonly known as the *hydraulic head*.

Figure 2-5 shows these heads diagrammatically superimposed on Figure 2-4. Note that as the velocity increases, so the pressure decreases – a conclusion that is

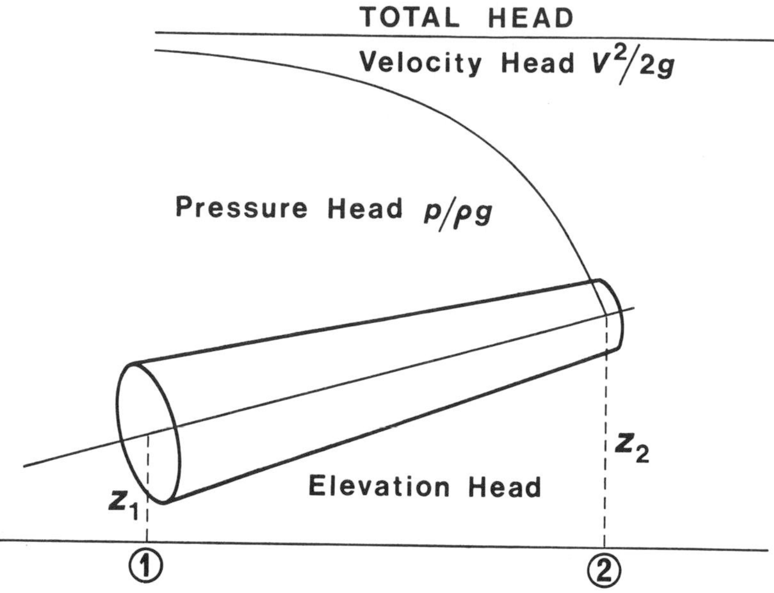

Fig. 2-5. Heads superimposed on Figure 2-4.

probably contrary to the intuitive, but experimentally verifiable and the basis of the venturi flow meter.

Real liquids, of course, are not frictionless; and in motion they lose energy in the form of heat generated by friction within the liquid. So with a real liquid, the total head decreases in the direction of flow, and equation (2.3) becomes

$$\frac{p_1}{\rho g} + z_1 + \frac{q_1^2}{2g} = \frac{p_2}{\rho g} + z_2 + \frac{q_2^2}{2g} + h_L, \tag{2.3a}$$

where h_L is the head loss incurred between stations 1 and 2.

We revert now to the matter of flow down an open channel of constant slope θ and uniform cross-section. If the liquid in the channel were an ideal liquid, without viscosity, without friction, Bernouilli's theorem tells us that the total head would remain constant. The elevation head is decreasing as sin θ. The component of the force due to the weight of the liquid along the channel, $\rho g \sin \theta$, induces as acceleration $g \sin \theta$ per unit of mass of the liquid, and the velocity head increases as the square of the velocity. The velocity of an ideal liquid in the channel increases indefinitely (there will come a time when the rate of increase of velocity head exceeds the rate of decrease of the elevation head) and the pressure head will correspondingly decrease. In other words, the liquid will become shallower in the channel in the direction of motion.

With water, a real liquid, in the channel, the water will accelerate until the resisting frictional force equals the gravitational force inducing acceleration, and thereafter the water moves at a constant velocity (by which we mean that the volume of water passing any transverse plane in unit time is constant and equal). From this point (Fig. 2-6) the total head decreases as the elevation head because both pressure and velocity heads remain constant. This loss of total head represents the energy dissipated, and is irrecoverable. It is equal to the loss of elevation head and is proportional to the frictional forces, or viscosity: it is a measure of the departure of the real liquid from an ideal liquid *in that channel*. The shape of the channel is a factor because the larger the ratio of water volume to wetted surface area, the more closely the water approaches the behaviour of an ideal liquid.

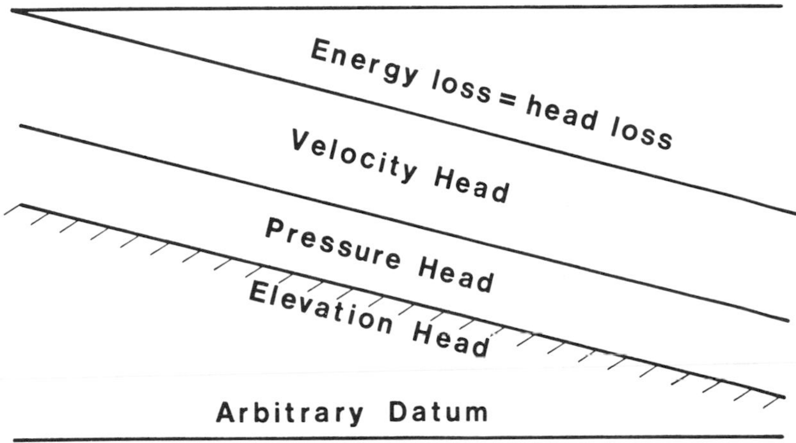

Fig. 2-6. Heads of uniform steady flow in open channel.

Channel-flow equations in common engineering use are not very satisfying (they are, of course, satisfactory or they would not be in common use). Probably the most commonly used equation for *turbulent* flow in channels and rivers is the Manning equation, which in its metric form is

$$q = \frac{1}{n} R^{\frac{2}{3}} S^{\frac{1}{2}}, \tag{2.4}$$

where q is the mean velocity, R the hydraulic radius, and $S = \Delta h/l$ is the slope of the water surface. n is a roughness coefficient known variously as Manning's or Kutter's n (these values are not altered according to the system of measurement used, but the numerical factor is, being 1 for metric, 1.49 for foot-pound-second units). This formula works well, but it is unsatisfying having and empirical equation with a coefficient having dimensions $T/L^{\frac{1}{3}}$, and no sign of g, ρ and η, which are hidden in there somewhere.

The role of these hidden components can be indicated by taking the earliest empirical channel-flow formula due to Chézy (in an unpublished note of 1775):

$$q = c\sqrt{RS}. \tag{2.5}$$

Now, c can be shown to be equal to $\sqrt{(8g/f)}$, where f is known as the friction factor. Also, for laminar flow, $f \propto \eta/Rq\rho$ (in other words, roughness is not a factor, in laminar flow: see, for example, Ackers, 1958, p. 2). Substituting these terms into Chézy's formula,

$$q = \left(\frac{C\rho g R q}{\eta}\right)^{\frac{1}{2}} (RS)^{\frac{1}{2}} = \left(\frac{C\rho g R^2 q S}{\eta}\right)^{\frac{1}{2}}$$

from which

$$q = C R^2 \frac{\rho}{\eta} g \frac{\Delta h}{l}. \tag{2.6}$$

Note carefully that equation (2.6) is valid for laminar flow only, and also that it is dimensionally homegeneous. It could have been derived from dimensional analysis, and the reader may care to try it.

We shall leave channel flow with the question, what would be the law of laminar flow through a channel filled with sand?

PIPE FLOW

Equations seeking to describe the flow of liquids in pipes are associated with the names of Poiseuille, a French doctor interested in blood flow through veins and arteries (1840, 1841), and Darcy, a French engineer employed in municipal water supply (1858), and others. It will better serve our purpose here to consult the work of Osborne Reynolds (1884, 1901), a British professor of Engineering, because his experiments covered the same topic, and introduced important new ideas.

Reynolds was aware that there are two types of flow in pipes, laminar and turbulent (or, as he put it, direct and sinuous), and he was well aware that the transition from one type of flow to the other depends on the velocity of the water in the pipe, the diameter of the pipe, and the viscosity of the liquid. He sought the criterion that determines whether flow is turbulent or laminar, and reasoned that since there is no measure of absolute time or absolute length, any length must be relative to some other length, and speed, to some other speed. He noted that the ratio of dynamic viscosity to mass density, η/ρ, has the dimensions of a length multiplied by a speed or velocity. He inferred that the criterion would be some critical value of the ratio of the products of two lengths and velocities, $Dq/(\eta/\rho) = Dq\rho/\eta$, where q is the mean velocity of the water (as before, volumetric discharge rate divided by cross-sectional area) in a pipe of diameter D, and $\eta/\rho = \nu$ is the kinematic viscosity of the liquid. The number is dimensionless, of course, so it does not matter which system of units is used provided it is consistent.

Figure 2-7a shows the mean velocity plotted against the hydraulic gradient as obtained from Reynolds' experiments with a lead pipe of 0.615 cm diameter (Reynolds, 1884, Table IV). The striking feature of the plot is the linear relationship between the hydraulic gradient (friction losses) and the mean velocity, with a change of slope at a velocity of about 0.54 m s^{-1}. Figure 2-7b shows the lower-velocity data of his Tables IV and V plotted logarithmically. This technique (introduced by Reynolds some years earlier) puts the data in the form $\log x = a \log y + \log b$, corresponding to $x = b\,y^a$. The slope of the log-log plot in Figure 2-7b indicates that at low velocities the exponent is close to unity, while the slope of the turbulent data is close to $\frac{1}{2}$. In other words, the friction losses are roughly

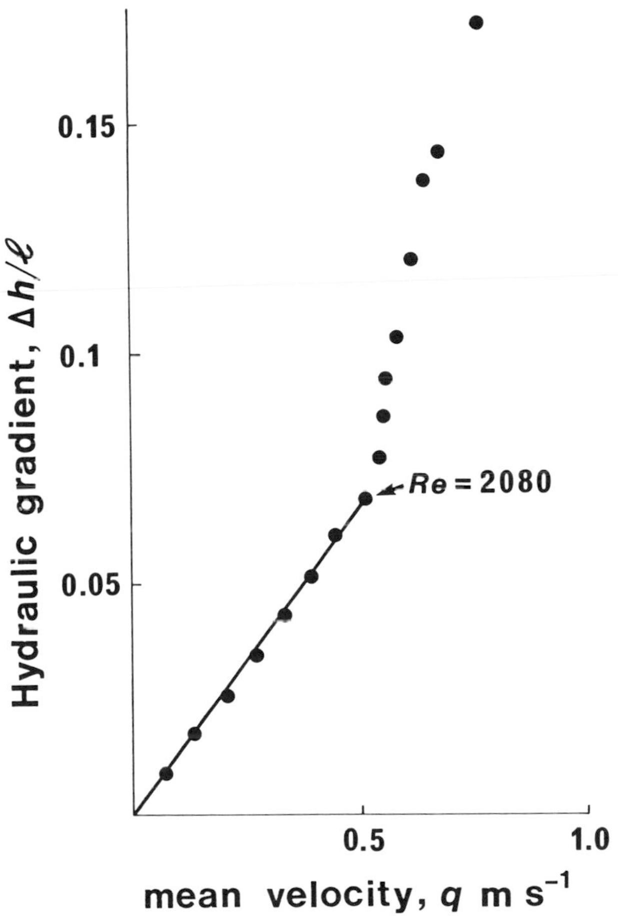

Fig. 2-7a. Relationship between mean velocity and hydraulic gradient found by Reynolds for a pipe of 0.615 cm diameter (Reynolds, 1884, Table IV).

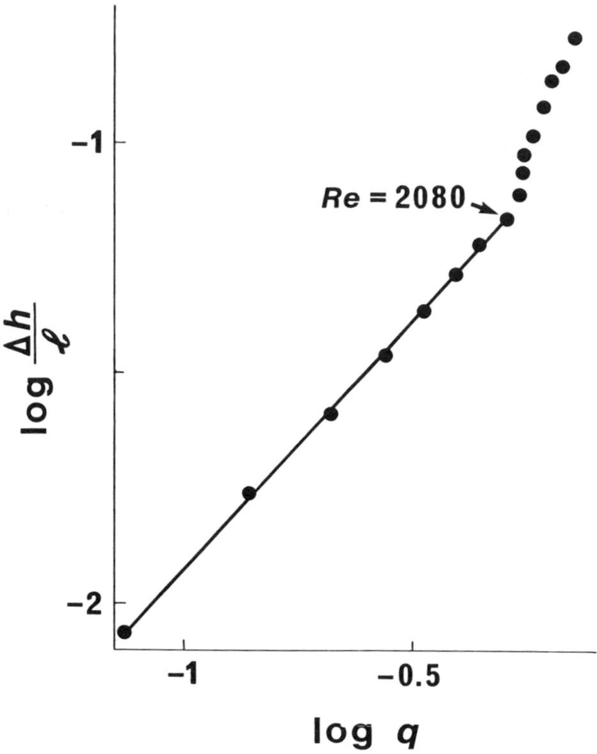

Fig. 2-7b. Reynolds' data of Figure 2-7a plotted as logarithms.

proportional to the mean velocity up to some critical mean velocity, and then, after a transition zone, roughly, proportional to the square of the mean velocity*.

Reynolds determined from his own experiments and those of Poiseuille and Darcy, that the velocity criterion is

$$q = K\eta/D\rho, \tag{2.7}$$

where K is a constant with a value about 2000 for pipes with circular cross-section. Turbulent flow occurs when $D\rho q/\eta > \sim 2000$ when q is the mean velocity and D is the diameter of the pipe. (This is not the number in his 1884 paper: he used a relative viscosity in that work and so found another number. His 1884 experimental data indicate a 'real critical value', as he called it, rather higher than 2000. His famous experiments with a coloured dye indicated a critical number averaging nearly 13,000 in 29 experiments!)

* The raw data in Reynolds' Tables III, IV, and V, is not linear in laminar flow, but this is because the water was cooling during this part of the experiments. When the viscosity is corrected for temperature, the linear relationship is found.

The number found by evaluating $D\rho q/\eta$ has come to be called *a Reynolds number*, designated *Re* or N_R. Note most carefully the indefinite article – *a* Reynolds number. If D is taken to be the hydraulic radius of the pipe ($\pi r^2/2\pi r = r/2 = D/4$) rather than its diameter, the critical Reynolds number has another value (~ 500), but the custom of writing $4R$ for D leads to the same number. A characteristic velocity and a characteristic dimension of length must be identified or chosen, and recorded.

Dimensional analysis helps us to an expression for the friction losses incurred during laminar flow of liquids in pipes. Reynolds' experiments indicate that the head loss is a function of kinematic viscosity ($v = \eta/\rho$), the mean velocity ($q = Q/A$), and a linear dimension characterizing the pipe. Bernouilli's theorem also indicates the presence of g, the acceleration due to gravity. We therefore postulate that the head loss incurred during laminar flow of water through pipe length l may be expressed.

$$\Delta h/l = C\, D^{a}\, q^{b}\, v^{c}\, g^{d},$$

where C is a dimensionless coefficient. Dimensionally,

$$LL^{-1} = (L)^{a}\,(LT^{-1})^{b}\,(L^{2}T^{-1})^{c}\,(LT^{-2})^{d}.$$

According to Buckingham's Π theorem (see any text book on fluid mechanics, and Buckingham, 1914, 1921) these five physical quantities expressed in two fundamental dimensions can be arranged into $5 - 2 = 3$ dimensionless groups, of which LL^{-1} is one, from which the form of the equation can be derived. Each of these dimensionless groups is called a Π-term. We take D and q as the dominant quantities, as Reynolds found, and make the two remaining Π-terms from (D, q, v) and (D, q, g).

We know that a Reynolds number can be formed from the first:

$$L^{0}T^{0} = (L)^{a}\,(LT^{-1})^{b}\,(L^{2}T^{-1})^{c}.$$

The indicial equations are,
for L: $0 = a + b + 2c$
for T: $0 = \quad - b - c$
from which $a = -c$, and $b = -c$, and the dimensionless number or group is $v/Dq = \Pi_2$. (This is, in fact, the ratio Reynolds used in 1884, not the inverse.)

The third group is formed from D, q, and g:

$$L^{0}T^{0} = (L)^{a}\,(LT^{-1})^{b}\,(LT^{-2})^{d}.$$

The indicial equations are,
for L: $0 = a + b + d$
for T: $0 = \quad - b - 2d$
from which $a = d$, $b = -2d$, and the dimensionless group is $Dg/q^2 = \Pi_3$. So,

$$\psi\,(\Pi_1, \Pi_2, \Pi_3) = \psi\,(\Delta h/l, \frac{v}{Dq}, \frac{Dg}{q^2}) = 0$$

and

$$\Delta h/l = \frac{v}{Dg}\,\psi_2\left(\frac{Dg}{q^2}\right).$$

Reynolds found that the hydraulic gradient in laminar flow was proportional to the velocity, hence

$$\Delta h/l = C\left(\frac{v}{Dq}\right)\left(\frac{q^2}{Dg}\right) = C\frac{qv}{D^2g}. \tag{2.8}$$

Analysis and experiment indicate that C has the *constant* value of 32 for pipes with circular section and diameter D, and that for laminar flow the nature of the pipe does not affect the frictional losses. The constant found for Reynolds' experiments in his Table IV is 32.8, a discrepancy well within the accuracy of the experiment because it could arise from a temperature error of less than 1 °C.

The equation

$$\Delta h/l = 32qv/gD^2 \tag{2.8a}$$

is known as the Hagen-Poiseuille, or the Poiseuille law for laminar flow in pipes*. It is the basis of much work on the flow of fluids through porous media. This equation can, of course, be re-arranged

$$q = \frac{D^2}{32}\frac{\rho}{\eta}\,g\,\frac{\Delta h}{l} \tag{2.8b}$$

and generalized by substituting $4R = D$,

$$q = \frac{R^2}{2}\frac{\rho}{\eta}\,g\,\frac{\Delta h}{l}. \tag{2.8c}$$

Note that this equation is in the same form as that for laminar flow in open channels (Eq. 2.6), and they could even be identical.

* It is worth noting in passing (although we shall not use it) that the hydraulic gradient can also be expressed as a function of velocity head, $q^2/2g$. Multiplying top and bottom of equation (2.8a) by $q/2$, we obtain

$$\Delta h/l = \frac{64q^2v}{2gD^2q} = \frac{64\eta}{\rho Dq}\frac{1}{D}\frac{q^2}{2g} = \frac{64}{Re}\frac{1}{D}\frac{q^2}{2g}.$$

This is the *Darcy-Weisbach* equation for laminar flow, the more general form being

$$\Delta h/l = f\frac{1}{D}\frac{q^2}{2g}$$

where f is here the *friction factor*.

DIMENSIONLESS NUMBERS

We have seen that a Reynolds number, $D\rho q/\eta$, is dimensionless (and intentionally so); and in the dimensional analysis above, two other dimensionless numbers were found, $\Delta h/l$ and Dg/q^2. The first of these is the hydraulic gradient. The inverse of the last, q^2/Dg, is known as *a Froude number* (Fr) – also its square root. A Froude number is important in naval architecture, but it is usually regarded as irrelevant in pipe-flow, or any other type of flow in which a free upper surface to the liquid is not evident. Nevertheless, the analysis revealed it, and it is not irrelevant.

Reynolds numbers are rationalized as the ratio of inertial to viscous forces: Froude numbers, as the ratio of inertial to gravitational forces. In our problems, a Froude number appeared because a free surface is implied when pressures are measured as manometer heights (or, in channel flow, because there is a free surface). In any case, all these flows take place under the force of gravity. If laminar flow is limited by a real critical value of a Reynolds number, it follows that in pipe flow there is also a real critical value of the corresponding Froude number, but this is not significant.

There are two great practical advantages of writing equations in terms of dimensionless numbers: any consistent system of measurement may be used, and, their use simplifies the experimental verification of the relationship. If measurements of pipe flow are made, and the results plotted as Reynolds number versus Froude number, the constant can be evaluated. In such experiments, it is only necessary to alter one of the variables in each number, and it does not matter which: so the simplest can be varied (in this case, the velocity and the hydraulic gradient, keeping the other quantities constant). This aspect can be pursued in any text-book on fluid mechanics.

PRESSURE

The reader will have noticed that *pressure* does not occur explicitly in any of the preceding flow equations (2.6, 2.8). Pressure is a force divided by an area on which the force acts, and it occurs only in the expression for pressure head (equation (2.3)). Pressure has the dimensions of (mass × acceleration)/area, that is, $MLT^{-2}L^{-2} = ML^{-1}T^{-2}$. Pressure head, on the other hand, has the dimension of *length*.

The whole subject of fluid movement has been plagued by confusion between these two quantities, and it is essential to keep the distinction between them clearly in mind. It is true that Poiseuille, Reynolds, Darcy, and countless later workers used the word *pressure* when they meant *pressure head*. The early workers' descriptions of their experiments leave no doubt about what they were meaning: this is not always true of later workers.

If, therefore, pressure is used instead of pressure head in any of the previous flow equations, they become physically erroneous because they become dimensionally unhomogeneous. For example, the Poiseuille law for laminar flow in pipes (equation (2.8b)) is commonly written

$$q = \frac{D^2}{32}\frac{g}{\eta}\frac{\Delta p}{l}.$$

This is *wrong*, and the reader who is not satisfied that it is wrong should not proceed until he is. It is wrong because it implies that when $q = 0$, $\Delta p = 0$. This is only true for a horizontal pipe. It is also dimensionally unhomogeneous.

As with channel flow, the question now arises, what is the flow equation for a pipe filled with sand? There will be little doubt that the form of the equation will be the same as those for channel and pipe flow, equations 2.6 and 2.8c. It is evident that the equation must take into account the greatly reduced area of flow (normal to the flow direction) and the greatly increased wetted perimeter (that is, the greatly reduced hydraulic radius). In other words, flow will be dominated by frictional resistance.

STOKES' LAW

In his investigations of the motion of pendulums, Stokes (1851; 1901, p. 60) developed analytically an expression for the terminal settling velocity of a solid sphere in a fluid *when the terms in V^2 can be neglected:*

$$V = \frac{2g}{9\nu}\left(\frac{\rho_s}{\rho_w} - 1\right)r^2 = \frac{gd^2}{18\nu}\left(\frac{\rho_s}{\rho_w} - 1\right), \tag{2.9}$$

where ν is the kinematic viscosity, ρ_s and ρ_w are the densities of the sphere and the fluid respectively, r is the radius and d the diameter of the sphere, and g the acceleration due to gravity. This equation is valid for any consistent set of units, but it is not valid when the Reynolds number, Vd/ν, exceeds one (approximately).

Stokes' law, as it has come to be called, must be used with these limitations in mind – a single solid particle settling at such a velocity that the Reynolds number does not exceed one. But the principle that settling velocity is determined partly by size and mass density is a useful one, and settling samples of sediment can lead to a rapid method of mechanical analysis, if the settling tube can be calibrated. There have been various attempts at doing this (for example, Rubey, 1933, Watson, 1969, and Gibbs et al., 1971) all meeting with some success. Dimensional analysis can help here.

We have seen that the six quantities, V, d, g, ρ_s, ρ_w, and ν, with the three dimensions of mass, length, and time, can be arranged into a function of three dimensionless groups or Π-numbers that are derived from the quantities involved. In this case, they can be seen to be

$\Pi_1 = Vd/v$, which is a Reynolds number;
$\Pi_2 = V^2/gd$, which is a Froude number; and
$\Pi_3 = (\rho_s/\rho_w) - 1$. (This form arises from the use of kinematic viscosity, rather than dynamic viscosity, and therefore a density term $(\rho_s - \rho_w)/\rho_w$.) Some combination of these three dimensionless numbers will give the form of the function.

One possibility is

$$\Pi_1 = b\,(\Pi_2/\Pi_3)^a, \tag{2.10}$$

which leads to Stokes' law when $a = 1$ *and* $b = 18$. But a and b must be constants, at least over a finite and useful range of Π_1, so that

$$\log \Pi_1 = a \log (\Pi_2/\Pi_3) + \log b \tag{2.11}$$

is the equation of a straight line. If experimental data in this form plots as one or more straight lines, the constants a and b can be evaluated by regression analysis on the data of each straight line, and all the equations are of the form of equation (2.10), which can be expanded and re-arranged into

$$V = \left\{\left(\frac{\rho_s}{\rho_w} - 1\right)^a g^a d^{a+1}/b\,v\right\}^{1/(2a-1)} \tag{2.12a}$$

and

$$d = \left\{b\,V^{2a-1}\left(\frac{\rho_s}{\rho_w} - 1\right)^{-a} g^{-a} v\right\}^{1/(a+1)}. \tag{2.12b}$$

When $a = 1$ and $b = 18$, Stokes' law for single solid spheres results.

Riley and Bryant (1979) calibrated their settling tube for analysis of multi-grain samples using industrial quartz spheres with diameters ranging from 74 to 710 μm, sieved to $\frac{1}{4}$-phi intervals. Several samples, each weighing 1 to 5 grammes, from each $\frac{1}{4}$-phi interval were settled. The grains landed on a tray to which was attached a strain gauge, by means of which a continuous recording of cumulative weight against time was obtained. The median velocity of the sample was then computed, and the geometric mean (the mid-point of the $\frac{1}{4}$-phi interval) was taken as the characteristic dimension of the sample. The water was at 20 °C, giving a kinematic viscosity of 1.007×10^{-6} m^2 s^{-1}. The density of the spheres was assumed to be 2,650 kg m^{-3}.

Computing the three Π-numbers from Riley and Bryant's data, we find that the logarithm of Π_1 versus the logarithm of Π_2/Π_3 plot as two straight lines (Fig. 2-8) with a change of slope at approximately $\log \Pi_1 = 0.37$ (Reynolds number, $Re = 2.4$). Regression analysis on the data of these two sets gives:

$$2 < Re < 60: a = 1.47, b = 55 \quad (r = 0.9987)$$
$$0.25 < Re < 2: a = 1.11, b = 25 \quad (r = 0.9956).$$

Table 2-1 compares the experimental data with the predictions of equations (2.12).

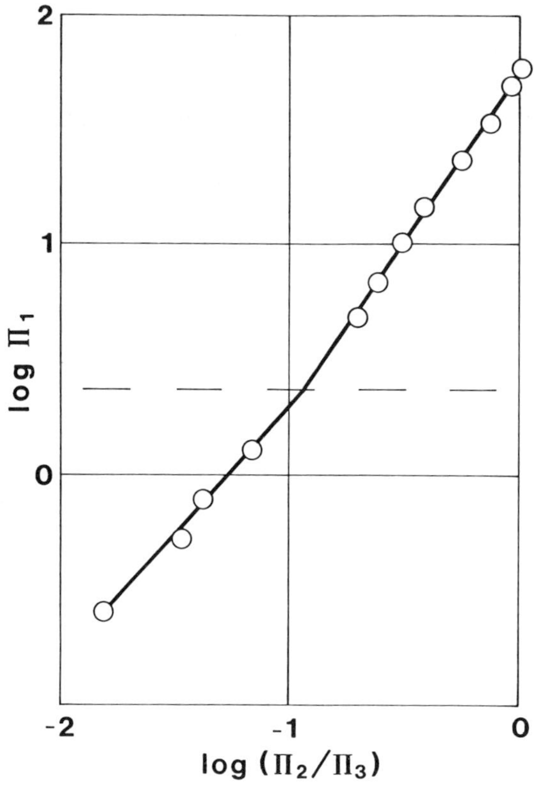

Fig. 2-8. Dimensionless plot of Riley and Bryant's (1979) settling tube calibration data for multi-sphere samples.

Table 2-1. Data of Riley and Bryant (1979) and predictions of equations (2.12).

V m s^{-1}	d μm	$\hat{V}$	$\hat{d}$
0.100	600	0.102	591
0.091	550	0.091	549
0.075	460	0.073	472
0.060	390	0.059	396
0.045	325	0.047	316
0.037	275	0.038	271
0.030	230	0.030	230
0.0250	195	0.0243	199
	$Re = 2$		
0.0113	115	0.0114	114
0.0081	98	0.0086	94
0.0066	81	0.0062	84
0.0040	64	0.0041	63

As we would expect with such high values of the correlation coefficient (r), the agreement is good. We conclude that equations (2.12) are valid calibration equations for the analysis of multi-grained samples in a settling tube, within the ranges of Reynolds number stipulated, and using the same technique of settling as in the calibration.

Gibbs et al. (1971) calibrated their settling tube for single spheres of various sizes and, in the larger sizes, different mass densities. The water in the tube was at 20 °C. Computing the three Π-numbers from their data, and plotting them as before (Fig. 2-9), we find a continuous curve that indicates that a and b are not constant. It is perhaps worth noting, though, that the curve can be closely approximated by four straight lines with changes of slope at $\log \Pi_1 = 0$, 1, and 2, corresponding with Reynolds numbers 1, 10, and 100. Regression analysis on these four sets of data gives us:

$$\begin{array}{llll} 100 < Re < 2300: & a = 3.20, & b = 40 & (r = 0.9964) \\ 10 < Re < 100: & a = 1.65, & b = 67 & (r = 0.9991) \\ 1 < Re < 10: & a = 1.21, & b = 38.6 & (r = 0.9988) \\ Re < 1: & a = 1, & b = 18 & \text{(Stokes' law).} \end{array}$$

Table 2-2 compares the observations with the predictions of equations (2.12), and again, we find good agreement, and can conclude that these equations are sufficiently accurate for single spheres in settling tubes, for the ranges of Reynolds number given above.

The values of a and b for multi-spheres are different from those for single solid spheres. This is not surprising (apart from the approximation used for single spheres). The significant difference between single-sphere settling and multi-sphere settling is the interaction between one grain and another, and the generation of sympathetic water movement in the tube. Clearly, the diameter of the tube in relation to the diameter of the particle or particles is also a parameter, because the larger the particle, the greater the interaction with the wall of the tube*.

Hydraulic equivalence

The sedimentological problem of assigning a diameter to grains of irregular shape is relatively simple if we take behaviour in water to be the guide. The hydraulic diameter is then the diameter of a sphere of the same material that has the same terminal settling velocity. The hydraulic mean diameter of a multi-grain sample is, by analogy, the mean diameter of a sample of spheres of the same material that has the same median settling velocity. (There are two troubles with this approach:

* There is a small device for measuring wind velocity that is based on this principle. A small, light, sphere is inside a narrow tube, the end of which is exposed to the wind. The stronger the wind, the higher the sphere is suspended. If the tube were of constant internal diameter, the device would not work: but it expands upward, and the wind velocity can be read against the scale opposite the small sphere.

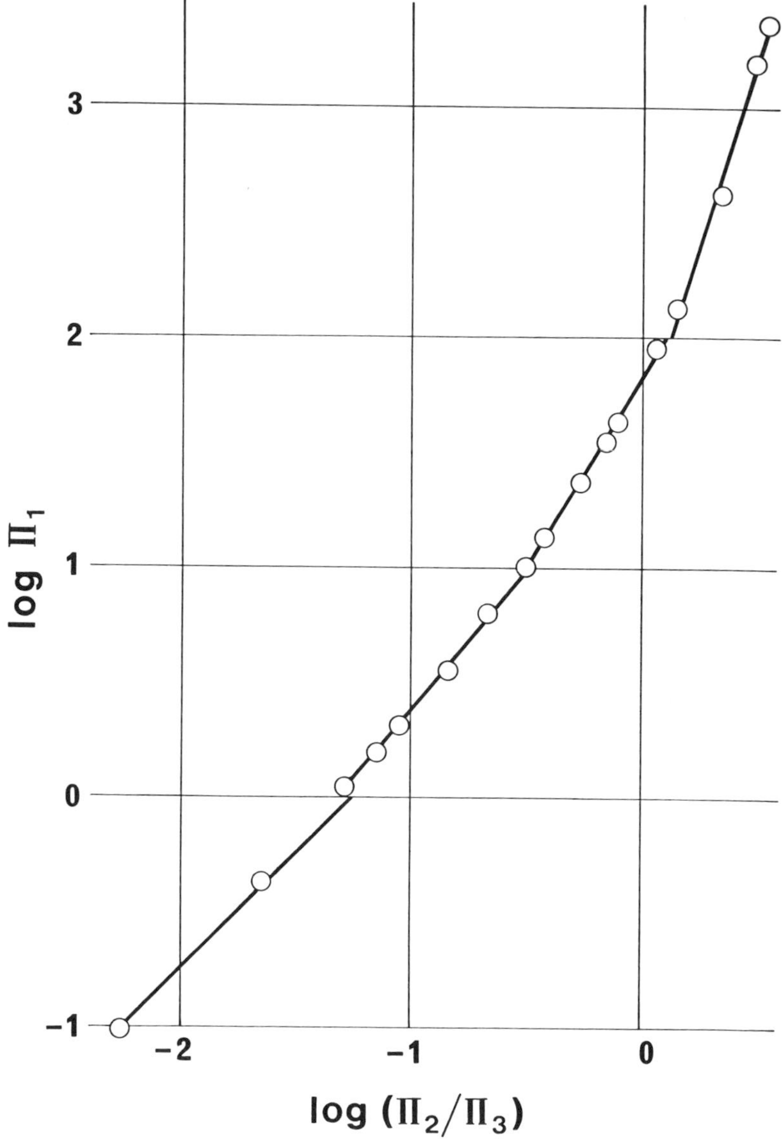

Fig. 2-9. Dimensionless plot of Gibbs et al.'s (1971) settling tube calibration data for single solid spheres.

grains of very irregular shape, such as needles, may have a preferred orientation while settling. And, more fundamentally, *settling from suspension* is not the usual process leading to accumulation of sediment in the stratigraphic record. But there are similarities between a grain's performance while settling, and when in traction along a horizontal surface, so the concepts are still useful.)

Equation (2.12b) gives the hydraulic diameter of a single grain, or the hydraulic

Table 2-2. Data of Gibbs et al. (1971) and predictions of equations (2.12) (single spheres).

V m s^{-1}	d μm		$\hat{V}$	$\hat{d}$
0.4607	5000		0.4644	4941
0.3904	4000		0.3904	3994
0.2507	1670		0.2432	1735
0.1400	950		0.1422	931
		$Re = 100$		
0.1148	783		0.1173	768
0.0790	550		0.0781	555
0.0711	500		0.0700	507
0.0572	417		0.0568	420
0.0421	323		0.0423	322
0.0359	283		0.0363	280
		$Re = 10$		
0.0271	233		0.0278	229
0.0196	183		0.0191	186
0.0139	150		0.0140	149
0.0118	133		0.0116	134
0.00940	118		0.00964	116
		$Re = 1$		
0.00521	84		0.00568	80
0.001977	50		0.002011	50

Note: the density of the first two spheres was 2,240 kg m^{-3}, the third, 2,755 kg m^{-3}, and the rest 2,488 kg m^{-3}.

mean diameter of a multi-grain sample, when its density (ρ_s) is inserted and the other quantities are known or measured.

There is an analogous concept: the diameter of a *quartz* sphere that would settle at the same terminal velocity as the grain of other material. In natural sediments, grains of heavy materials tend to accumulate with larger grains of quartz: this was called the 'equivalent hydraulic size' by Rittenhouse (1943). Equation (2.12a) leads to simple expressions for this relationship.

Equating the right-hand side of equation (2.12a) for the density of quartz with that for the material of density ρ_*, we obtain

$$d_*/d = \{(\frac{\rho_*}{\rho_w} - 1)/1.65\}^{a/(a+1)}, \qquad (2.13a)$$

where d_* is the hydraulic equivalent size, d is the hydraulic diameter of the material of density ρ_*, and the appropriate value of a is taken according to the Reynolds-number range (the exponent is not very sensitive, and will lie usually between 0.5 and 0.6). The ratio d_*/d is constant within each range of Reynolds numbers. If this is to be expressed in phi-units, the *difference* is constant:

$$\phi_* - \phi = \frac{3.32a}{a+1} \log_{10} \{1.65/(\frac{\rho_*}{\rho_w} - 1)\}. \qquad (2.13b)$$

The use of these equations is best illustrated by an example. A sample of rutile ($\rho_*/\rho_w = 4.2$) is found to have a median settling velocity of 0.052 m s^{-1} in water at 20° C ($\nu = 1.007 \times 10^{-6}$ m^2 s^{-1}). Inserting these values into equation (2.12b) gives an hydraulic mean diameter of 240 μm (rounded off; Reynolds number 12). Equation (2.13a) gives the ratio of hydraulic equivalent mean size to hydraulic mean diameter, $d_*/d = (3.2/1.65)^{0.6} = 1.5$. So the hydraulic equivalent mean size is $1.5 \times 240 = 360$ μm.

The use of dimensionless numbers in such work leads to simpler expressions than those derived by other methods, and there is the added advantage that not only may the settling tube be used at whatever temperature is convenient (provided it is known), but one is not restricted to water as the settling medium. Obviously there are limitations on the size of material to be analyzed, but the smaller sizes may be more readily analysed by using a liquid of smaller kinematic viscosity than water, thus speeding up the analysis without loss of accuracy. By the same token, material that would fall outside the range of calibration in water may be brought within it by changing the liquid.

Fluidization is a process in which solid material is supported in an upward flow of fluid, and it may be important in some aspects of volcanism. Strictly, it is the 'fluidizing' of an otherwise coherent porous solid, and so is a variety of Darcy's law rather than Stokes'. The principle is the same, although the Reynolds number may well be outside the realm of both laws.

NOTE ON POISEUILLE'S EXPERIMENTS

Poiseuille, in his studies of water flow through very small capillary tubes of diameters 0.013 to 0.65 *milli*metres, found an expression for what may be regarded as relative kinematic viscosity. His pipe flow equations (in his notation) were

$$Q = k'' \frac{PD^4}{L} \quad \text{and} \quad V = \frac{4k''}{\pi} \frac{PD^2}{L},$$

where Q is the discharge in cubic millimetres per second and V is the velocity in millimetres per second (Poiseuille, 1840, pp. 1046-1047). P is the *pression*, a manometer reading in millimetres corresponding to Δh because the discharge end of the capillaries was open to the atmosphere ('Nous avons emprunté la pression, non à la charge du liquide qui s'écoule, mais à un manomètre à air libre, soit à eau, soit à mercure . . .', op. cit., p. 962). Since P, D, and L all have dimensions of length (Poiseuille used mm), Q of volume/time, and V of velocity, k″ has dimensions $[L^{-1}T^{-1}]$, that is, $\rho g/\eta$ or g/ν (he realised that k″ included the density of the liquid).

In part IV of his work, Poiseuille (1841, pp. 112-115) found that the relationship between k″ and temperature T was

$$k'' = k_1 (1 + 0.033679T + 0.0002209936T^2)$$

where k_1 is the value of k'' at 0°C, which he found to be 135.28 ($mm^{-1}\ sec^{-1}$) for water manometers.

Comparing Poiseuille's equation for velocity with equation (2-8b), we find

$$\frac{4k''}{\pi}\frac{PD^2}{L} = \frac{D^2}{32}\frac{\rho}{\eta} g \frac{\Delta h}{l}$$

from which

$$k'' = \frac{\pi}{4}\frac{g}{32\nu}.$$

So we can write

$$\nu = \frac{\pi}{4}\frac{g}{32k''} = \frac{\pi}{4}\frac{g}{32k_1}(1 + 0.033679T + 0.0002209936T^2)^{-1}\ (mm^2/sec^{-1})$$

or

$$\nu = 1.779 \times 10^{-6} (1 + 0.033679T + 0.0002209936T^2)^{-1}\ (m^2\ s^{-1}).$$

This formula for the kinematic viscosity of water at temperatures within the range of Poiseuille's experiments is highly satisfactory, as Table 2-3 shows.

Table 2-3. Comparison of kinematic viscosities ($m^2\ s^{-1}$) predicted by Poiseuille's formula (column 1) with values commonly accepted today (column 2).

T °C	(1)	(2)
10	1.31×10^{-6}	1.31×10^{-6}
20	1.01×10^{-6}	1.01×10^{-6}
30	8.05×10^{-7}	8.04×10^{-7}
40	6.59×10^{-7}	6.61×10^{-7}
45	6.00×10^{-7}	6.05×10^{-7}

SELECTED BIBLIOGRAPHY

Ackers, P., 1958. Resistance of fluids flowing in channels and pipes. *Dept. Scientific and Industrial Research, London, Hydraulics Research Paper* 1 (39 pp.).

Buckingham, E., 1914. On physically similar systems; illustrations of the use of dimensional equations. *Physical Review*, 2nd series, 4: 345-376.

Buckingham, E., 1921. Notes on the method of dimensions. *Lond. Edinb. Phil. Dubl. Mag.*, 6th series, 42: 696-719.

Darcy, H., 1858. Recherches expérimentales relatives au mouvement de l'eau dans les tuyaux. *Mémoires présentés par divers savants à l'Académie des Sciences de l'Institut de France* 2e serie, 15: 141-403.

Emery, K.O., 1938. Rapid method of mechanical analysis of sands. *J. Sedimentary Petrology*, 8: 105-111.

Gibbs, R.J., Matthews, M.D., and Link, D.A., 1971. The relationship between sphere size and settling velocity. *J. Sedimentary Petrology*, 41: 7-18.

Poiseuille, J.L.M., 1840. Recherches expérimentales sur le mouvement des liquides dans les tubes de très petits diamètres. *Comptes rendus hebdomadaires des Séances de l'Académie des Sciences, Paris*, 11: 961-967, 1041-1048.

Poiseuille, J.L.M., 1841. Recherches expérimentales sur le mouvement des liquides dans les tubes de très petits diamètres. IV. Influence de la temperature sur la quantité de liquide qui traverse les tubes de très petits diamètres. *Comptes rendus hebdomadaires des Séances de l'Académie des Sciences, Paris*, 12: 112-115.

Reynolds, O., 1884. An experimental investigation of the circumstances which determine whether the motion of water shall be direct or sinuous, and the law of resistance in parallel channels. *Philosophical Trans. Royal Society*, 174 (for 1883): 935-982.

Reynolds, O., 1901. *Papers on mechanical and physical subjects, Volume II* (1881-1900). Cambridge University Press, Cambridge, 740 pp.

Riley, S.J., and Bryant, T., 1979. The relationship between settling velocity and grainsize values. *J. Geol. Soc. Australia*, 26: 313-315.

Rittenhouse, G., 1943. Transportation and deposition of heavy minerals. *Bull. Geol. Soc. America*, 54 (12): 1725-1780.

Rubey, W.W., 1933. Settling velocities of gravel, sand, and silt particles. *American J. Science*, 25: 325-338.

Stokes, G.G., 1851. On the effect of the internal friction of fluids on the motion of pendulums. *Trans. Cambridge Philosophical Soc.*, 9: 8-106. (Not sighted.)

Stokes, G.G., 1901. On the effect of the internal friction of fluids on the motion of pendulums. *In*: G.G. Stokes, *Mathematical and physical papers, Volume* 3. Cambridge University Press, Cambridge, pp. 1-141.

Taylor, E.H., 1939. Velocity-distribution in open channels. *Trans. American Geophysical Union*, 20: 641-643.

Vanoni, V.A., 1941. Velocity distribution in open channels. *Civil Engineering*, 11 (6): 356-357.

Watson, R.L., 1969. Modified Rubey's law accurately predicts sediment settling velocities. *Water Resources Research*, 5: 1147-1150.

3. LIQUID FLOW THROUGH POROUS SANDS

We have considered so far those aspects of fluid statics and fluid dynamics that are essential for an understanding of the main theme of this book – pore water and geology – and we have seen that the topics can barely be considered without some mathematics, but that the mathematics could be done by a High School student. Mathematics is a language that more geologists understand nowadays, but it is worth remembering that much of applied mathematics in the natural sciences is either trivial or intractable. It is not our purpose here to derive the laws of flow through porous solids from the Navier-Stokes equation, but rather to express them in terms of the measurable and useful parameters of sedimentary rocks. Our purpose is to understand the processes rather than to develop predictive equations or formulae. Those who seek a higher goal may find the note at the end of the chapter a useful starting point.

HENRY DARCY'S EXPERIMENTS

Henry Darcy was an engineer concerned with the public water supply to the town of Dijon, France, and in 1856 he published as an appendix to a broader report the results of experiments conducted 'to determine the laws of water flow through sands' (Darcy, 1856, pp. 590-594; reprinted in Hubbert, 1969, pp. 303-311. See also Hubbert, 1940, p. 787 et seq., 1956 and 1957, for analyses of Darcy's experiments).

The experimental apparatus is shown diagrammatically in Figure 3-1. It consists of a cylinder in which there was a measured column of sand supported by a filter. Water was passed through the cylinder, and the volumetric rate of flow measured at the outlet. Manometers were placed near the top and the bottom of the sand (Darcy used mercury manometers, and computed the readings that would have been obtained with water manometers). At various rates of flow, the difference in elevation between the mercury levels in the two manometers was measured, and computed for water. The results of three series of experiments are plotted in Figure 3-2.

Darcy observed that 'for sands of the same nature' the flow of water was proportional to the difference in elevation between the levels in the manometers (as computed for water) and inversely proportional to the length of the sand in the

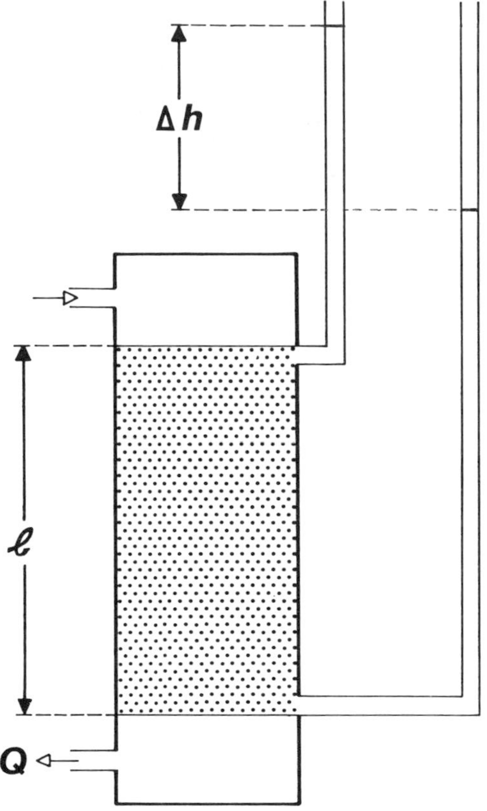

Fig. 3-1. Diagram of Darcy's apparatus.

direction of water flow. From his observations he deduced (in the notation used in this book)

$$Q = K\frac{A}{l}(h_1 + l \pm h_2), \tag{3.1}$$

where Q is the volume of flow in unit time, l is the length of sand traversed, A is the gross cross-sectional area normal to the flow, h_1 is the computed or equivalent water height in the inlet manometer above the top of the sand, h_2 is the computed water height in the outlet manometer above (−) or below (+) the base of the supporting filter, and K is 'a coefficient depending on the permeability of the bed' of sand.

We must pause here to note with some surprise that not all the experiments Darcy reported support the linear expression he deduced. If a straight line is drawn on Figure 3-2, 1st Series, with the constraint that it passes through the origin, and the slope adjusted so that as many points fall above the line as below it, the first five points fall below the line and the second five above it. Figure 3-3 shows Darcy's data

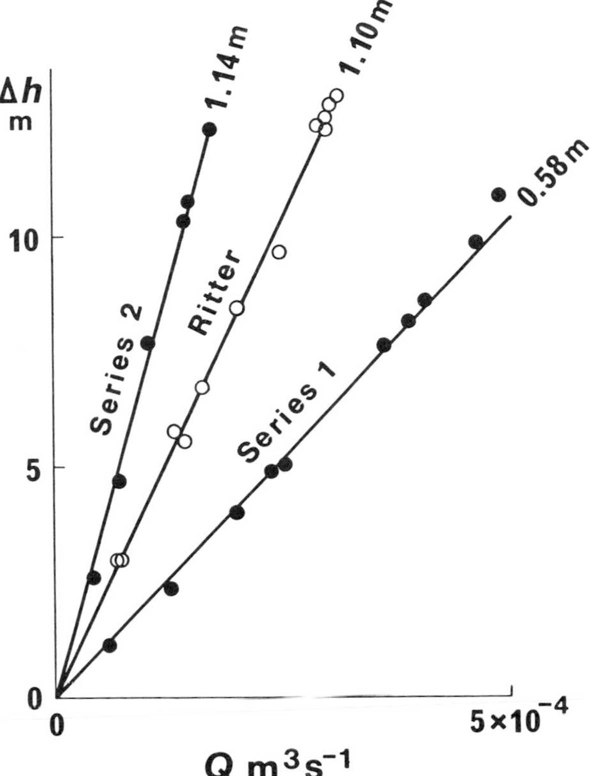

Fig. 3-2. Results of three of Darcy's experiments.

(first two series, and Ritter's results) plotted on logarithmic scales; only the series carried out by M. Ritter shows a strictly linear relationship (and the 4th Series, with three points, which is not shown), the rest having slopes less than unity (between 0.87 and 0.92). Darcy probably relied on Ritter's experiments, and the linear relationship has since been amply demonstrated, so we must guess that the water was cooling down during the first three series of experiments (as indeed it was for Reynold's experiments, with the same result).

It is critically important for the understanding of Darcy's experiments to note that '*Toutes les pressions ont été rapportées au niveau de la face inférieure du filtre*' (*op. cit*., p. 592, my italics) – 'All pressures have been referred to the lower surface of the filter' – and that he was therefore measuring what we would now call the sum of the elevation and pressure heads. This is quite explicit in his equation in the term $h_1 + l$ because l is the elevation head (in his experiments) above the supporting filter and h_1 is the pressure head above the top of the sand. *He was not measuring the pressure difference across the sand.*

In the conventional modern notation, Darcy's law is written

$$q = Q/A = -K\frac{\Delta h}{l}, \tag{3.2}$$

where q is the *specific discharge* (or *discharge velocity*) and K is Darcy's coefficient of proportionality that depends on the permeability of the porous material. The minus sign follows the convention that q is in the direction of decreasing total head (h); but it is commonly omitted (as we shall do) when the directon of flow can be obtained by inspection.

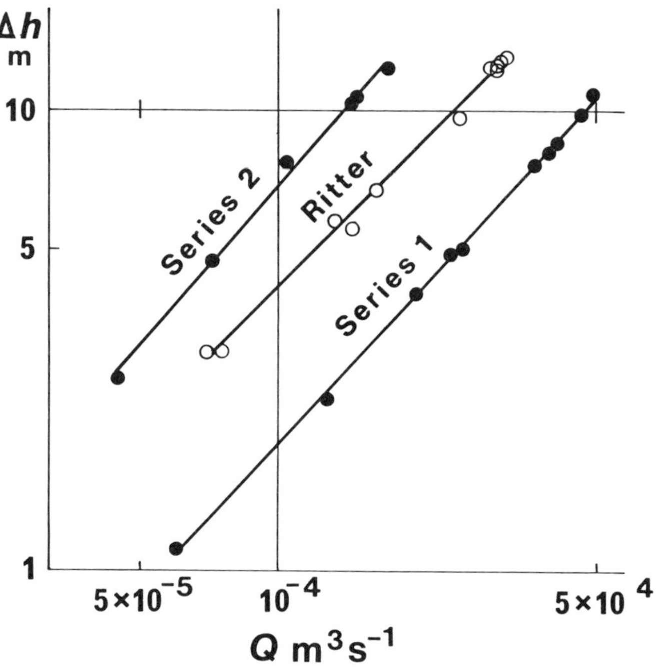

Fig. 3-3. Data of Figure 3-2 plotted on logarithmic scales.

THE COEFFICIENT OF PERMEABILITY

The coefficient of permeability, K, clearly takes several influences into account. Darcy used water for his experiments, but had he used oil or any other liquid that is inert to the solids he would no doubt have obtained quantitatively different results. He did obtain different results with different sands. The coefficient K therefore consists of at least two parts: one related to the material alone, the other to the liquid.

The dimensions of K are $[L^3T^{-1}]\,[L^{-2}]\,[LL^{-1}] = LT^{-1}$, which are the dimensions of a velocity. Since $\Delta h/l$ is dimensionless, q also has the dimensions of a velocity, but it is a notional velocity in the sand because only part of the cross-sectional area is available for liquid movement. K therefore includes this effect, but we shall consider it later. The obvious physical properties of the liquid are (from the considerations of Chapter 2) its mass density ρ, with dimensions ML^{-3}, and dynamic viscosity η, with dimensions $ML^{-1}T^{-1}$. Gravity is clearly a driving force, so we include g, with dimensions LT^{-2} (Q could have been measured as a weight flow rate). There remains the significant dimension of the permeable material. Many workers (e.g., Lindquist, 1933; Hubbert, 1940, 1956, 1957; Schneebeli, 1955) have taken the characteristic dimension to be the mean grain diameter; but this is unlikely to be adequate because it is a very indirect measure of the geometry of the pore space through which the liquid passes. Let us try to clear this problem up.

For liquid flow in channels and pipes, as we have seen, the characteristic dimension is found to be the *hydraulic radius*

$$R = \frac{\text{area of transverse section of liquid}}{\text{wetted perimeter}}$$

$$= \frac{\text{volume of liquid}}{\text{wetted surface area}}. \tag{3.3}$$

This approach to flow through porous solids was taken by Blake (1923), Kozeny (1927a, 1927b) and Fair and Hatch (1933). The obvious difficulty is that part of the surface area of the grains (for example, close to grain contacts) may not exert any significant influence on the fluid flow; but the material constant should take this into account.

If we consider bulk volume v of the permeable material, the product of the effective porosity and the bulk volume, fv, is the volume of movable liquid. The wetted surface area of the grains in the bulk volume can be calculated for grains of simple geometry, leaving the contact areas to be taken into the material constant. If the grains are all spherical and of equal diameter, the surface area of each grain is πd^2. With n grains in bulk volume v, their total surface area is

$$S_n = n\,\pi\,d^2. \tag{3.4}$$

The total volume of these grains is

$$v_n = \frac{n\,\pi\,d^3}{6} \tag{3.5}$$

and they occupy $(1-f)$ of the bulk volume of the material. The bulk volume is therefore

$$v = \frac{n\,\pi\,d^3}{6(1-f)} \tag{3.6}$$

and the ratio of grain surface area to bulk volume, known as the *specific surface* (S) is

$$S = S_n/v = \frac{n\,\pi\,d^2}{(n\,\pi\,d^3)/6(1-f)} = \frac{6(1-f)}{d}. \tag{3.7}$$

The smaller the grain diameter, the larger the specific surface. For non-spherical grains, the factor will be larger than 6 because spheres have the smallest ratio of surface area to volume.

Our purpose being to understand the roles of the measurable and observable parameters of granular sedimentary rocks, not to develop predictive equations, we take the characteristic dimension of the pore space to be

$$R = f/S = fd/(1-f) \tag{3.8}$$

leaving the dimensionless coefficient to be taken into a general coefficient.

What meaning is to be attached to the mean grain diameter d when the grains are of unequal size? The mean diameter appears in the hydraulic radius in the term for the specific surface (equation (3-7)), so it is the contribution to the specific surface that is significant, and the interpretation of mean diameter must be that that gives the true specific surface.

Consider a sediment composed of two fractions only, w_1/W by weight of grains of diameter d_1, and w_2/W by weight of grains of diameter d_2. If the grains are geometrically similar, and of the same material or density, the total weight is proportional to the total grain volume $(1-f)$, and the weight of each fraction proportional to its grain volume. Following the argument in the derivation of the hydraulic radius (equations (3.4) to (3.7), the contribution to the specific surface from the d_1 fraction is

$$S_1 = \frac{c_1\,(1-f)}{d_1}\,\frac{w_1}{W} \tag{3.9}$$

and likewise, from the d_2 fraction

$$S_2 = \frac{c_1\,(1-f)}{d_2}\,\frac{w_2}{W}. \tag{3.9a}$$

There exists a mean diameter d such that

$$\frac{c_1\,(1-f)}{d} = \frac{c_1\,(1-f)w_1}{d_1 W} + \frac{c_1\,(1-f)w_2}{d_2 W} \tag{3.10}$$

or

$$\frac{1}{d} = \frac{1}{d_1}\,\frac{w_1}{W} + \frac{1}{d_2}\,\frac{w_2}{W}. \tag{3.11}$$

This is the *harmonic mean diameter* weighted by the weight fraction.Note that the harmonic mean takes size-sorting into account (to some extent at least) because the greater the dispersion (of positive numbers) the smaller the harmonic mean is relative to the arithmetic and geometric means.

Tortuosity

There is yet another property of the pore space that must be taken into account. It was noted earlier (as most previous investigators have noted) that the specific discharge q is a notional velocity in the porous material (it is the real velocity of the water above and below the sand in Darcy's experiments), so K contains a dimensionless factor that takes this into account. Only the effective pore space is available for water flow, so the mean water velocity in the sand is apparently q/f. The real velocity through the pore space is greater than that because the pore passages are not straight, and a particle of water must travel a mean distance l_t during its displacement over the linear length l of a sand (parallel to the macroscopic flow direction). This dimensionless property of a sedimentary rock is known generally as *tortuosity*, which we shall define $T = l_t/l$; it is difficult to define quantitatively and evaluate.

Consider a cube of impermeable material penetrated by a single non-linear capillary tube of circular cross-section that passes from one face to a similar position on the opposite face of the cube (Fig. 3-4). It is clear a) that the length of the capillary is greater than the distance between the faces, b) the area of intersection of the capillary with the bounding faces is greater than cross-sectional area of the capillary normal to the flow, and, on surfaces cut through the cube parallel to the bounding faces, the area of intersection with the capillary will only be equal to the cross-sectional area where the axis is normal to the surface; and c) that energy losses

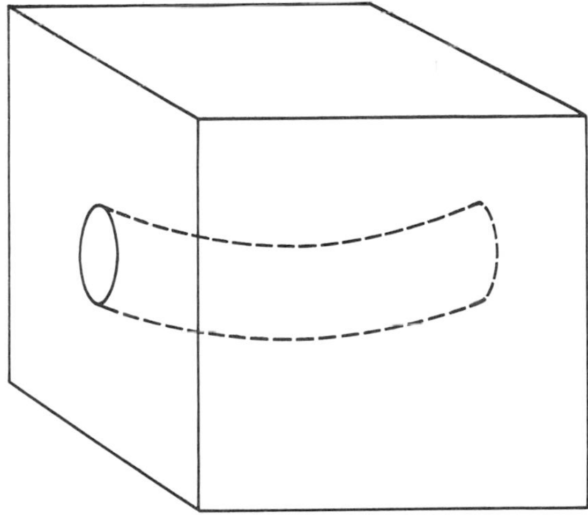

Fig. 3-4.

incurred in water flow through the capillary are greater than those incurred in flow through a straight capillary of the same diameter.

Now consider a cube of impermeable material of side l penetrated by n non-linear, non-intersecting, capillaries of diameter x, the mean length of which is l_t, the volume of the capillaries amounting to, say, 20% of the total volume of the cube. The capillaries pass from one face to the parallel face, the bounding faces being normal to the mean flow direction. We infer that

1) the mean inclination between the axes of the capillaries and the normal to the bounding faces, a, is given by $\cos a = l/l_t$.

2) the circular sections of the capillaries appear, in general, as ellipses on the bounding faces (and any surface parallel to them) with the minor axis equal to the diameter of the capillaries (x), and the major axis equal on average to $x/\cos a = xl_t/l$.

3) the surface porosity, defined as the proportion of pore space in the total area of a surface – in this case the bounding faces – is therefore

$$f_s = \frac{n\pi x^2 l_t}{4l} \times \frac{1}{l^2}. \tag{3.12}$$

This is seen to be also the bulk porosity f due to capillaries of diameter x and mean length l_t in a cube of volume l^3.

There are several obvious defects in this model. In real porous materials, the capillaries are not of constant or equal diameter, they intersect one another, and they pass in all directions through the cube (although there may be a preferred orientation in anisotropic material). But however we refine the model, it seems that the term fA exaggerates the area of pores available for water flow by a factor of l_t/l because of the 'mean inclination' of the pore channels. So the expression we suggested for the real velocity (q/f) should be ql_t/fl. The real pore velocity, as we have seen, is greater than this by a factor l_t/l, so the real pore velocity V_p is related to q by

$$V_p = \frac{q}{f}\left(\frac{l_t}{l}\right)^2. \tag{3.13}$$

Since q is *defined* by Q/A, the coefficient of permeability K contains a factor $f(l/l_t)^2$.

We are still not finished with tortuosity, however, because the hydraulic gradient, $\Delta h/l$, is also affected by it: the gradient is exaggerated because $l < l_t$. So it also requires a factor l/l_t, which is also contained in K.

We conclude, therefore, that the total effect of tortuosity is

$$T^{-3} = \left(\frac{l}{l_t}\right)^3 \tag{3.14}$$

and that K contains the factor f as well as T^{-3}.

Reverting now to the components of the coefficient of permeability, K, and the expansion of Darcy's law, dimensional analysis of the variables with dimensions (ρ, η, g, R) leads to an expression similar to Equation 2.6; and, with f and T,

$$K = C f T^{-3} R^2 \frac{\rho}{\eta} g \tag{3.15}$$

and Darcy's law can now be written

$$q = C f T^{-3} R^2 \frac{\rho}{\eta} g \frac{\Delta h}{l}. \tag{3.15a}$$

For the evaluation of tortuosity we turn to the electrical measurement of resistivity of porous solids saturated with an electrolyte, in particular to Archie's Formation Resistivity Factor. Archie (1942), following on earlier work, found that the resistivity of a porous solid, R_0, is related to the resistivity of the electrolyte saturating the pore space, R_w, by

$$R_0 = F R_w, \tag{3.16}$$

where F is a dimensionless material constant known as the Formation Resistivity Factor, or just Formation Factor.

By an argument strictly analogous to the one just used for permeability, the Formation Factor can be expressed in terms of porosity, capillary lengths and cross-sectional areas:

$$F = R_0/R_w = \frac{R_w \times l_t}{fl} \frac{l_t}{l} \frac{1}{R_w} = \frac{l_t^2}{fl^2}. \tag{3.17}$$

Following Street (1958) we see that

$$(Ff)^{-1.5} = (l/l_t)^3 = T^{-3}. \tag{3.17a}$$

But Archie suggested that the data supported the relationship $F = f^{-m}$, where m is a number between about 1.3 for loose, unconsolidated sands to about 2.5 for consolidated sands. So

$$\begin{aligned} f T^{-3} &= f(f^{-m} f)^{-1.5} \\ &= f^{1.5m - 0.5}. \end{aligned} \tag{3.18}$$

Substituting equation (3.18) into equation (3.15) we obtain an expansion of the coefficient of permeability in terms of all the foregoing considerations,

$$K = C f^{1.5m - 0.5} R^2 \frac{\rho}{\eta} g. \tag{3.19}$$

(Although $R = fd/(1-f)$, it will be more convenient for the time being to keep tortuosity separate from the hydraulic radius, which is a measure of pore size.)

Verification of this relationship is sought by plotting experimental data in dimensionless form, the dimensionless groups or numbers being derived from the variables in the manner used to obtain equation (2.8) in the previous chapter.

The variables are $K, fT^{-3}, R, \rho, \eta, g$. Six variables with the three dimensions of M, L, Time, can be arranged into $(6 - 3 =)$ three dimensionless groups of which fT^{-3}, being dimensionless, is one (Π_1). Choosing K, ρ, η, as the repeating variables common to both the other groups, and including R in one and g in the other, we find

$$M^0 L^0 T^0 = (LT^{-1})^a (ML^{-3})^b (ML^{-1}T^{-1})^c (L)^d$$

and the indicial equations are,

for M: $\quad 0 = \quad b + c$

for L: $\quad 0 = a - 3b - c + d$

fot $Time$: $\quad 0 = -a \quad - c$

from which, $b = -c = a = d$. Since a Π term can be replaced by any power of that term,

$$\Pi_2 = \frac{K\rho R}{\eta}$$

which will be recognized as a Reynolds number with K a velocity. For Π_3 we use K, ρ, n and g, obtaining

$$\Pi_3 = \frac{K^3\rho}{g\eta}.$$

These three groups are consistent with equations (3.15) and (3.19) if we write

$$\Pi_3/\Pi_1 = C\,\Pi_2^2.$$

Π_3 is found to be the product of the Reynolds number Π_2 and a Froude number (the ratio of inertial to gravity forces)

$$(K\rho R/\eta)\,(K^2/gR) = K^3\rho/\eta g\,.$$

So writing Fr for K^2T^3/gRf since Π_3/Π_1 is also a Froude number,

$$Re \times Fr = C\,Re^2$$

from which

$$Fr = C\,Re$$

or

$$\left(\frac{K^2T^3}{gRf}\right) = C\left(\frac{K\rho R}{\eta}\right). \tag{3.20}$$

The dimensionless groups are thus found to be those in brackets in equation (3.20).

Schriever (1930) carried out some careful experiments on the permeability of packings of glass spheres to hot oil, within the realm of Darcy's law (for there is a realm, as we shall see). Each of four series of experiments was made with spheres of a single diameter, but the porosity was varied by altering the packing by hammering. He thus obtained measurements of the permeability of each of four sizes for four different porosities. Schriever's paper is of no theoretical interest now: he was seeking to evaluate the constants of Slichter's (1899) equation, which has been abandoned in favour of those by Kozeny (1927) or Fair and Hatch (1933) or variants of these.

Taking the cross-sectional area used in his formula on p. 335 of his paper, and evaluating fT^{-3} by assigning m the value 1.3 for loose sands, his data have been tabulated in Table 3-1 and plotted in Figure 3-5 in dimensionless form, log Re versus log Fr, revealing the relationship

$$\log Fr = a \log Re + \log C$$

corresponding to $Fr = C\,Re^{a}$. The slope of the line indicates that $a = 1$; and the value of C (computed from Schriever's data) is found to be 8.42×10^{-3}. The number C is dimensionless, but it is not a true constant because it contains at least one coefficient relating to the shape of the grains (in the expression for hydraulic radius).

Table 3-1.Grain size, porosity, and permeability. (Data of Schriever, 1930, p. 335, Table 1)

d cm	f	K cm/s	$R = fd/(1-f)$	$Re = \dfrac{K\rho R}{\eta}$	$Fr = \dfrac{K^2}{gRf^{1\cdot45}}$	$C = Fr/Re$
0.1025	0.3870	1.4411×10^{-1}	6.4710×10^{-2}	1.5592×10^{-1}	1.2959×10^{-3}	8.3×10^{-3}
	0.3777	1.2657	6.2212	1.3166	1.0771	8.2
	0.3653	1.0986	5.8994	1.0836	8.9817×10^{-4}	8.3
	0.3533	9.5533×10^{-2}	5.5997	8.9445×10^{-2}	7.5104	8.4
0.0528	0.3889	3.8869×10^{-2}	3.3602×10^{-2}	2.1838×10^{-2}	1.8026×10^{-4}	8.3×10^{-3}
	0.3779	3.5590	3.2074	1.9086	1.6506	8.6
	0.3689	3.1120	3.0863	1.6059	1.3582	8.5
	0.3603	2.8071	2.9739	1.3958	1.1867	8.5
0.0443	0.3958	3.0159×10^{-2}	2.9020×10^{-2}	1.4634×10^{-2}	1.2250×10^{-4}	8.4×10^{-3}
	0.3849	2.6525	2.7721	1.2294	1.0329	8.4
	0.3715	2.3087	2.6185	1.0108	8.7211×10^{-5}	8.6
	0.3552	1.8463	2.4403	7.5332×10^{-3}	6.3871	8.5
0.0252	0.3934	9.7454×10^{-3}	1.6343×10^{-2}	2.6630×10^{-3}	2.2913×10^{-5}	8.6×10^{-3}
	0.38055	8.0996	1.5481	2.0965	1.7533	8.4
	0.3690	7.0929	1.4737	1.7477	1.4770	8.5
	0.3597	6.2742	1.4157	1.4851	1.2484	8.4

Notes: 1) First two columns, Schriever's data. Third column computed from his data by dividing by the area A and the ratio of mercury density to his oil density ($14.77 \times 13.55/0.836 = 239.39$). The area was obtained from his formula on p. 335 and the data of his table (β is correctly given in spite of the error of sign).
2) $\rho/\eta = 16.72$ s/cm^2. The value assigned to m is 1.3 for loose sand, hence the exponent 1.45.
3) Dimensionless numbers are the same in any consistent set of units: Schriever used *c.g.s.*, and the same result is obtained with SI units.

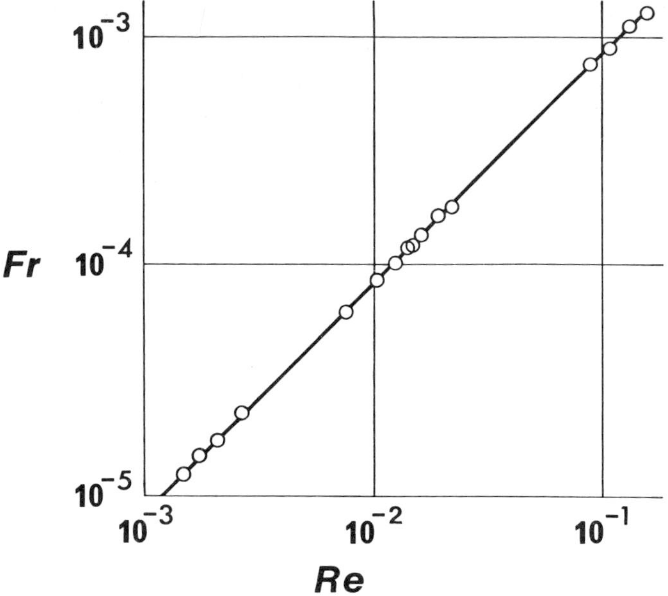

Fig. 3-5. Dimensionless plot of Schriever's (1930) data (Table 3-1).

We therefore write Darcy's law

$$q = Q/A = K\frac{\Delta h}{l}$$

$$= CfT^{-3}\,R^2\frac{\rho}{\eta}g\frac{\Delta h}{l}. \qquad (3.21)$$

where $fT^{-3} = f^{1.5m-0.5}$, where m is the same 'cementation factor' used in Archie's Formation Factor. And reverting to Equation 3.19, we see that the combined porosity term is

$$f^{1.5m+1.5}/(1-f)^2 = f^x/(1-f)^2,$$

where x varies from 3.5 to 5.3 as m varies from 1.3 to 2.5.

The coefficient of proportionality K, the *coefficient of permeability*, is also commonly called the *hydraulic conductivity*. We noted early on that this coefficient consists of at least two parts, one due to the material alone, another due to the liquid. The term $CfT^{-3}\,R^2 = Cf^x d^2/(1-f)^2$ relates solely to the porous material; it has dimensions L^2, and this quantity is called the *intrinsic permeability*, with the symbol k:

$$K = k\frac{\rho g}{\eta}.$$

The term ρ/η, the inverse of the kinematic viscosity, relates to the liquid (not necessarily to gases). The last two terms, $g(\Delta h/l)$, relate to the energy loss in the system. We have already seen (pp. 30 to 32) that the hydraulic gradient, $\Delta h/l$, can be expanded by Bernouilli's theorem in terms of *pressure head*, *elevation head* and *velocity head*. So the complete expansion of Darcy's law, for our purposes, is

$$q = CfT^{-3} R^2 \frac{\rho}{\eta} g \frac{1}{l} \left[\frac{p_1 - p_2}{\rho g} + (z_1 - z_2) \right]$$

$$= C \frac{f^x d^2}{(1-f)^2} \frac{\rho}{\eta} g \frac{1}{l} \left[\frac{p_1 - p_2}{\rho g} + (z_1 - z_2) \right], \qquad (3.22)$$

where the suffixes refer to two points along the macroscopic flow line separated by distance l measured also along the macroscopic flow line. The velocity head has been omitted because it is negligibly small, being the difference of two very small quantities compared to the pressure and elevation heads.

Let us now review the terms in this expansion of Darcy's law, because it is important to understand each.

- The specific discharge q is a notional velocity across a plane surface normal to the macroscopic flow direction. It has a real value, of course, and it is the basis for calculating ground water flow rates and well yields. It has the dimensions of length divided by time, but the length is more readily understood as a volume divided by an area.
- The coefficient C is dimensionless, but it is not a true constant because it includes a coefficient from the hydraulic radius, the value of which in C lies between $1/6^2$ for spheres and $1/8^2$ approximately. There is also a coefficient from Archie's relationship $F = bf^{-m}$, but this coefficient b is close to unity, varying from about 0.6 to 1.3 in sands (see Keller, 1966, p. 563, Table 26-5). We shall examine the coefficient C again before the end of the chapter.
- The tortuosity $T = l_t/l$, takes into account the non-linear pore paths, the mean inclination of which to the macroscopic flow direction is given by $\cos a = 1/T$. From equation (3.18), $T^{-3} = f^{1.5m-1.5}$, so T varies as $f^{0.5-0.5m}$ where m varies from 1.3 to about 2.5. In other words, tortuosity is a function of porosity, and for 20% porosity, T varies from 1.3 to about 3.3 and a varies from 40° to 70°.

The use of Archie's Formation Factor for the evaluation of tortuosity depends on the assumption of strict analogy between the flow of electicity and liquids. There are cogent arguments in support of this, for Hubbert has shown that Darcy's law and Ohm's law are strictly analogous both physically and mathematically (Hubbert, 1940, p. 819; see also Versluys, 1930, p. 217).

Many sedimentary rocks are anisotropic with respect to permeability (and, indeed, to electrical resistivity). Of particular significance are the vertical and horizontal permeabilities, the latter normally being greater than the former. It is evident that the expression for intrinsic permeability must include a vector quantity, and the only vector quantity appears to be tortousity.

The exponent m is called the *cementation factor*, and the name has been criticized as inappropriate. The main reason why ordinary cementation seems to have little effect on permeability (Füchtbauer, 1967, p. 359) seems to be that it first forms pendular rings around the points of contact, which is the volume of pore space that contributes least to flow because of the velocity gradient to the static boundary surface (Fig. 3-6) (see also Rose, 1959). It seems, therefore, that m is a factor that takes *pore shape* into account, for porosities less than about 26% are impossible with packings of spheres of similar sizes. For straight capillaries parallel to the macroscopic flow direction, $m = 1$, and the tortuosity is equal to 1.

– The hydraulic radius $R = fd/(1-f)$ takes the wetted surface area into account, and is a function of both porosity and the *harmonic mean* grain diameter. The harmonic mean diameter can be computed from the same data from which the geometric mean is calculated, and with the same limitations (such as using the bounding sieve sizes to obtain a measure of the mean size of the grains retained on the smaller sieve).

The role of porosity in permeability is seen to be a composite effect due to tortuosity, hydraulic radius, and the first correction to the notional velocity q. Permeability varies as $f^{1.5m-0.5}f^2/(1-f)^2 = f^{1.5m+1.5}/(1-f)^2$. As m varies from 1.3 to 2.5, permeability varies as $f^{3.5}/(1-f)^2$ to $f^{5.3}/(1-f)^2$.

– The kinematic viscosity $v = \eta/\rho$ inversely affects the hydraulic conductivity K: the more viscous and less dense the liquid, the smaller the specific discharge for a constant hydraulic gradient through a given material. Kinematic viscosity has the dimensions of a length times a velocity, hence the units are $m^2 s^{-1}$ in the SI system, where the mass density is in $kg\, m^{-3}$ and the dynamic viscosity is in $kg\, m^{-1} s^{-1}$ (the dimensions also reveal why these two viscosities are called kinematic and dynamic).

– The hydraulic gradient $\Delta h/l$ is commonly misunderstood and misrepresented, although its significance has long been recognized (Norton, 1897, Plate VIII and p. 173, drew an 'isopiestic' map of Iowa, with sea-level as datum, and used the term 'hydraulic gradient'). The total head h is the algebraic sum of the pressure head $p/\rho g$ and the elevation head z: it has the dimension of length (as all heads do). Versluys (1917) called this the 'hydrostatic potential' or simply 'potential', and mentioned that we are usually concerned with potential differences. The hydraulic gradient is

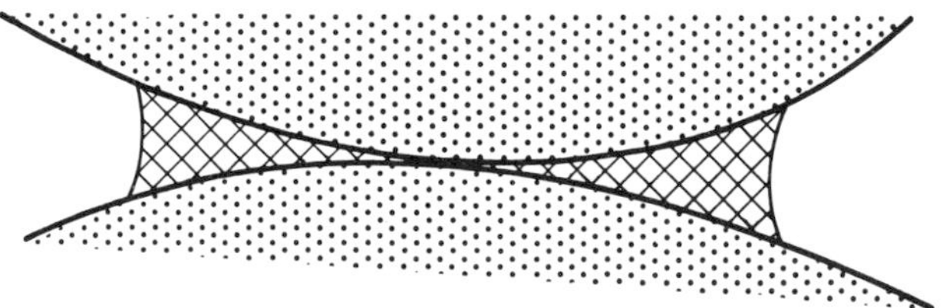

Fig. 3-6. Pendular cement occupies space that would have contributed little to liquid flow.

the gradient of total head along the macroscopic flow path of length l: it is dimensionless. Hubbert (1969, p. 15) drew attention to some early misunderstandings, but the position is now so confused that it would probably be better if a whole new terminology could be devised. The 1972 AGI Glossary of Geology is wrong in defining hydraulic gradient as 'the rate of change of *pressure head* per unit of distance of flow' (their italics). It is wrong because static pore water in an inclined aquifer has a pressure-head gradient but no flow. This error arises from their definition of pressure head, in which they wrongly refer it 'to a specific level such as land surface'. Pressure head only makes sense when it is measured from the aquifer: the introduction of an arbitrary datum immediately changes this to total head, which is correctly defined in the 1972 AGI Glossary of Geology. Figure 3-7 makes these distinctions clear.

The total head is an energy per unit of weight: the product gh is an energy per unit of mass, with dimensions L^2T^{-2}. Hubbert (1940, pp. 796-803), in an analysis that is much more rigorous and general than that here, called gh the *potential* of the fluid, with the symbol Φ. So the term $g\Delta h/l$ is the potential gradient.

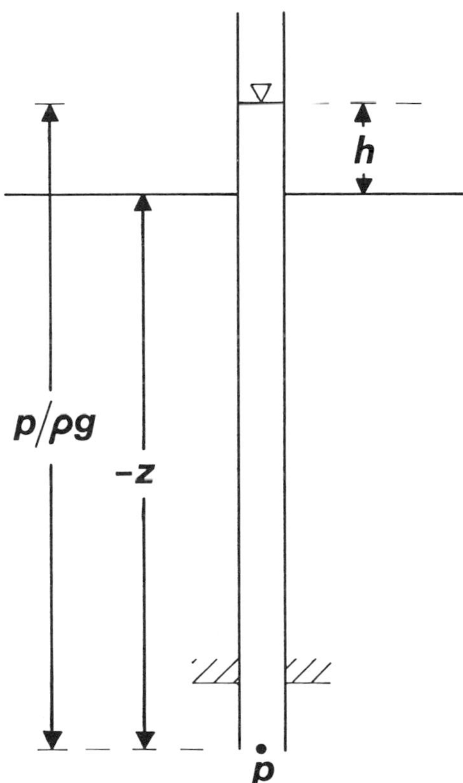

Fig. 3-7. Relationship between elevation head (z), pressure head ($p/\rho g$), and total head (h).

THE LIMITS OF DARCY'S LAW

The existence of a lower limit to Darcy's law has been postulated on the grounds of liquid adsorption to solid surfaces, molecular and pore-throat sizes. Whether this is really so on a practical scale is hard to determine, and in any case depends on what we mean by a statement that liquid flow 'follows Darcy's law'. If we mean that there is a linear relationship between the energy losses and the specific discharge, then Darcy's law holds for permeable glass (Fig. 3-8) with an intrinsic permeability of the order of 10^{-16} cm^2 or 10^{-8} darcies (see Glossary) for water, acetone, and *n*-decane (the permeabilities being computed from data reported by Nordberg, 1944, and Debye and Cleland, 1959). This is near the lower limit of intrinsic permeabilities measured in shales as reported by Bredehoeft and Hanshaw (1968) and Magara (1971).

The matter is very complex because meticulous experimental techniques are required for the measurement of such small permeabilities. Olsen (1965, 1966) concluded that Darcy's law is valid for most clays, and that the deviations from Darcy's law observed by Hansbo (1960), Lutz and Kemper (1963), and von Engelhardt and Tunn (1955) could be attributed to the experimental conditions, but that some of the threshold gradients for fluid flow found by Miller and Low (1963) were larger than the possible experimental error. (It should be noted that Miller and Low found a linear relationship for clays with about 70% porosity, non-linear for those with about 85% porosity – well beyond the limits of sedimentary rocks.)

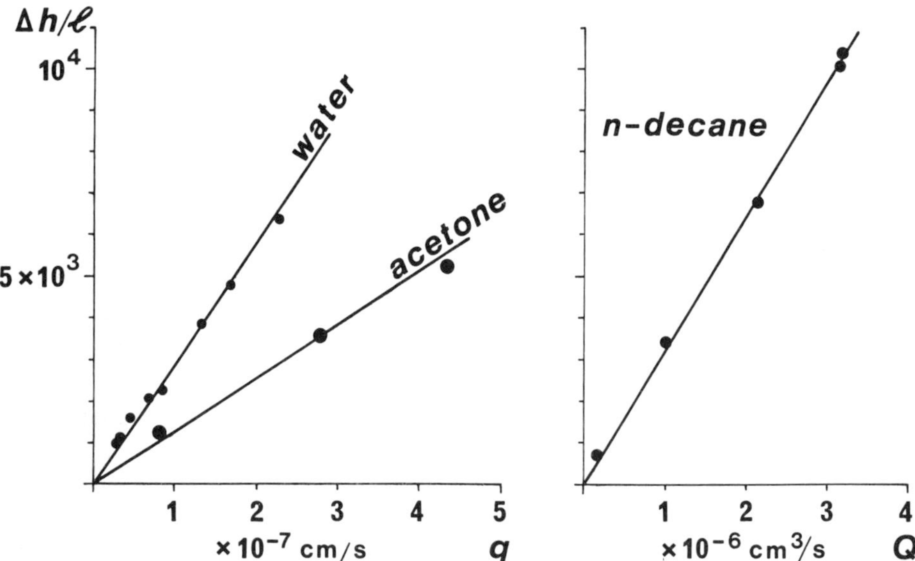

Fig. 3-8. Data of Nordberg (1944) on left, and Debye and Cleland (1959) show that Darcy's law applies to permeable glass.

Nordberg's data (Fig. 3-8) possibly indicates a threshold hydraulic gradient of about 100 for liquid flow, a small intercept at $q = 0$ being indicated. Debye and Cleland took evaporation into account, and no significant intercept is found. So evaporation could account for Nordberg's intercept (that is, apparent $q < 0$ at $\Delta h = 0$ rather than $\Delta h > 0$ at $q = 0$).

It is interesting to compare the values of the constant C found from Schriever's (1930) and Nordberg's (1944) experiments. Expressing the hydraulic radius more explicitly by $R^* = f/S$ where S is the specific surface, the expression for intrinsic permeability may be written

$$k = C^* fT^{-3} \frac{f^2}{S^2}. \tag{3.23}$$

Nordberg reports the following data for VYCOR, the brand of porous glass he used: $f = 0.28$, $S = 1.9 \times 10^6$ cm^2/cm^3. The tortuosity $T = l_t/l$ of another sample of VYCOR is about 2.5 (Barrer and Barrie, 1952). The mean value of k from Nordberg's data is found to be 2.65×10^{-16} cm^2; hence $C^* = 0.7$.

The value of C found from Schriever's experiments (8.42×10^{-3}) included a shape factor from the hydraulic radius $(\frac{1}{6})^2$; so for comparative purposes, Schriever's $C^* = 0.3$. The intrinsic permeability of Schriever's material varied from 9×10^{-6} cm^2 to 4×10^{-7} cm^2. The value of C^* from the experiments of Schriever and of Nordberg are remarkably close in view of the great disparity of intrinsic permeabilities, Schriever's material being of the order of 10^9 to 10^{10} more permeable than Nordberg's porous glass. Could the true value be 0.5, and Darcy's law for laminar flow of liquids through porous solids identical with that for laminar flow of liquids through pipes (Equation 2.8c)?

Upper limit of Darcy's law

There is an upper limit to Darcy's law. For a given material, there is a value of q, the specific discharge, above which the energy losses cease to be linear with q and increase more rapidly. In other words, there is an upper limit to the range in which the coefficient of permeability K is constant. It is customary to relate this limit to a Reynolds number based on q and the mean grain diameter d (rarely defined, but geometric mean sometimes implied) – and in these terms, the upper limit of Darcy's law is found to fall at a Reynolds number between 1 and 10. The actual range found may be greater than that*, but this need not concern us greatly because a Reynolds number based on q and d in this way is hardly likely to be satisfactory. The notional

* Not as great as that reported by Scheidegger (1960, p. 159). The low value of 0.1 (Nielsen, 1951) is based on a different Reynolds number from the high value of 75 (Plain and Morrison, 1954) – both being different from the $Re = q\rho d/\eta$ used by most workers. Nielsen's Reynolds number is not readily converted; but Plain and Morrison's 75 corresponds with a Reynolds number $q\rho d/\eta \simeq 15$. However, their data do not permit reliable conclusions to be drawn concerning the true upper limit of Darcy's law.

velocity q is related to the mean pore velocity by the factor fT^{-2} and the mean grain diameter is a characteristic length of that part of the material through which the fluid does not pass (it has been compared to taking the pipe thickness as the characteristic dimension of pipe flow – unfairly, because the grain diameter is a measure of pore size, even if a poor one). A more realistic Reynolds number for fluid flow through porous solids would be

$$\frac{qT^2}{f}\,\frac{\rho}{\eta}\,\frac{fd}{a(1-\mathrm{f})},$$

where d is the harmonic mean grain diameter and a takes the value 6 for spherical grains, to about 8 for irregular shapes. Using the argument leading to Equations 3.17 and 3.18, we see that $fT^{-2}=f^m$, so there is a porosity factor in this Reynolds number, $f^{1-m}/a(1-f)$. If $f=0.3$, $m=1.3$, and $a=7$, the porosity factor is 0.3. Reynolds number based on q and mean grain diameter are too large, and do not take all the significant parameters of the porous solid into account.

Lindquist (1933) made an interesting study of water flow through porous soils and came to the conclusion that Darcy's law failed at a Reynolds number about 4 (based on q and mean diameter), but that flow was still laminar to $Re=180$; and that the failure of Darcy's law is due to inertial forces in the flow through irregular pore spaces (Lindquist, 1933, pp. 89-91). Bakhmeteff and Feodoroff (1937) also found Darcy's law to fail at a Reynolds number about 4, and they clearly understood the nature of this failure.

Hubbert (1940, pp. 819-822) has emphasized that the first failure of Darcy's law is due not to turbulence but to the inertial forces' becoming significant. Brownell and Katz (1947), when allowance is made for their porosity term, also found a critical Reynolds number about 5 above which Darcy's law did not apply; they thought that this failure was due to turbulence in some paths, not in others.

Schneebeli (1955) found that Darcy's law failed at a Reynolds number about 5 for spheres, 2 for granite chips, but that turbulence set in at a Reynolds number about 60 for both materials. These results suggest that shape *may* influence the point at which Darcy's law fails, as well as tortuosity; but the difference could be due to porosity. The effect of porosity is shown in Figure 3-9, in which Schriever's data is plotted against Schneebeli's dimensionless numbers.*

* Statistics is no substitute for thought. Regression analysis of Schneebeli's groups with Schriever's data suggests the relationship $C_f=1433\,Re^{-0.99}$ with $r=-0.994$ and a level of significance $<0.1\%$. From this one might conclude that $C_f\times Re=$ *constant*, and so $d^2/K=$ *constant* for constant porosity. However, regression analysis of the data of each diameter leads consistently to a (log/log) slope of -2.000 and $r=-1$. For constant diameter, $C_f \propto Re^{-2}$ from which we find $gd^3(\rho/\eta)^2=$ *constant*. This tells us nothing we did not already know, because all the other quantities were constant in Schriever's experiments.

Two valid conclusions can be drawn: Schriever's experimental techniques were meticulous, and at least one pertinent dimensionless variable is missing from Schneebeli's groups. It is because there are only limited ranges of porosity to be obtained from porous solids that the suppression of porosity in the first result has little effect on the statistical significance.

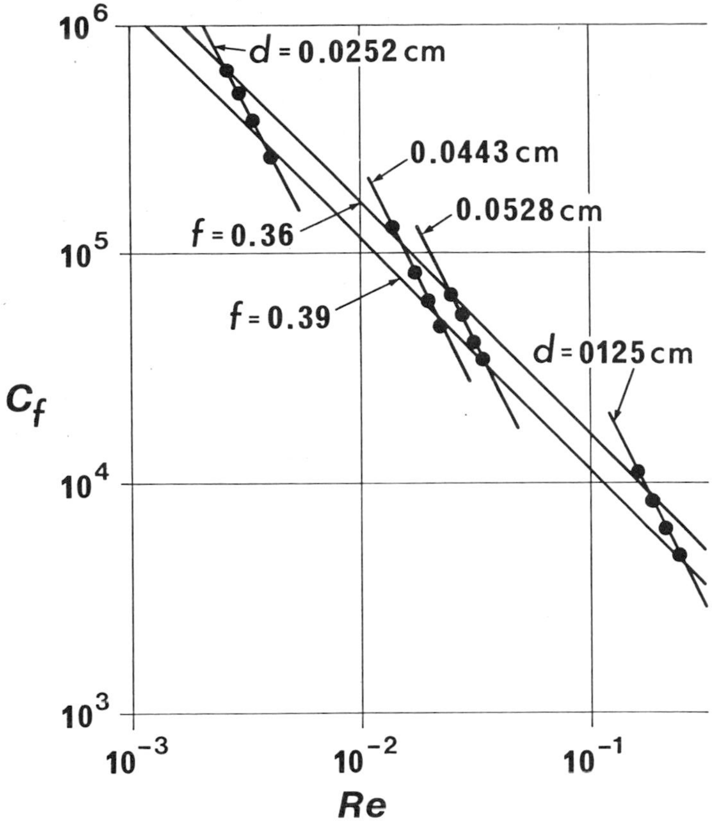

Fig. 3-9. Schriever's (1930) data plotted against Schneebeli's (1955) dimensionless groups.

Fortunately, most ground-water flow takes place within the realm of Darcy's law, and the real critical value of the Reynolds number that takes porosity into account, and used the harmonic mean grain diameter (always smaller than the geometric or arithmetic mean) appears to be close to 1.

The generalization of Darcy's law for liquids is an oversimplification in the geological context; but it helps us to understand the more complex reality, the meaning of permeability, and the factors that affect it. Real sediments display variable grain size, variable sorting, porosity and tortuosity, and these may change along the flow path. The movement of subsurface water through porous sediments (as distinct from fissured rocks) may involve changing geometry of the bed as a whole, changing intrinsic permeability and, through changing temperatures, changing viscosity and density of the liquid.

Consider water flow through two layers of contrasting sand (Fig. 3-10) in apparatus similar to Darcy's. The conservation of matter requires that the volumetric

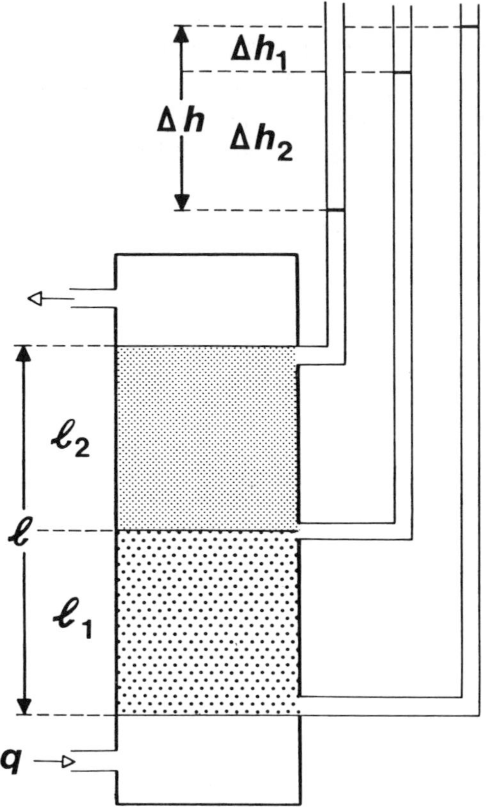

Fig. 3-10. Apparatus for investigation flow through two contrasting sands.

rate of flow per unit of cross-sectional area in the first layer shall be the same as that across the second layer. Hence,

$$Q/A = q = K_1 \frac{\rho}{\eta} g \frac{\Delta h_1}{l_1} = k_2 \frac{\rho}{\eta} g \frac{\Delta h_2}{l_2} = \bar{k} \frac{\rho}{\eta} g \frac{\Delta h}{l}. \tag{3.24}$$

If $\rho g/\eta$ can be regarded as constant, then

$$k_1/k_2 = \frac{\Delta h_2/l_2}{\Delta h_1/l_1} \tag{3.25}$$

and the ratio of the intrinsic permeabilities (and their hydraulic conductivities) of the layers is equal to the inverse ratio of their hydraulic gradients. If, for example, the hydraulic gradient across the first layer is half that across the second, the permeability of the second layer is half that of the first. The two layers together behave as if they were one layer with permeability $\bar{k}$ and hydraulic gradient $\Delta h/l$. Darcy's law, in practical terms, is a statistical relationship in which inhomogeneities are averaged out between manometer readings. The greater the number of measurements of total head along a flow path, the more detailed the information we can obtain about permeability changes. Pressure and total head measurements or estimates are more reliably obtained at present than *in situ* permeability measurements.

In the geological context, the geothermal gradient makes variables of the dynamic and kinematic viscosities, η and ν ($=\eta/\rho$). Strictly, pressure also affects these, increasing the density and increasing the viscosity of water at subsurface temperatures (see Bett and Cappi, 1965). Figure 3-11 shows the approximate change of ν under subsurface conditions of normal hydrostatic pressure and a geothermal gradient of 36°C/km. While these changes are important because increasing the temperature from 25°C to 100°C increases the hydraulic conductivity of a permeable rock by a factor of 3, such temperature changes will occur usually over a depth range of about 2 km. Vertical water movement during compaction will commonly be through less than 500 m, with a temperature difference of less than 20°C. At temperatures above 70°C, corresponding to depths below about 1,500 m, the change of viscosity is relatively slight (about 25% for a 20°C change) and no

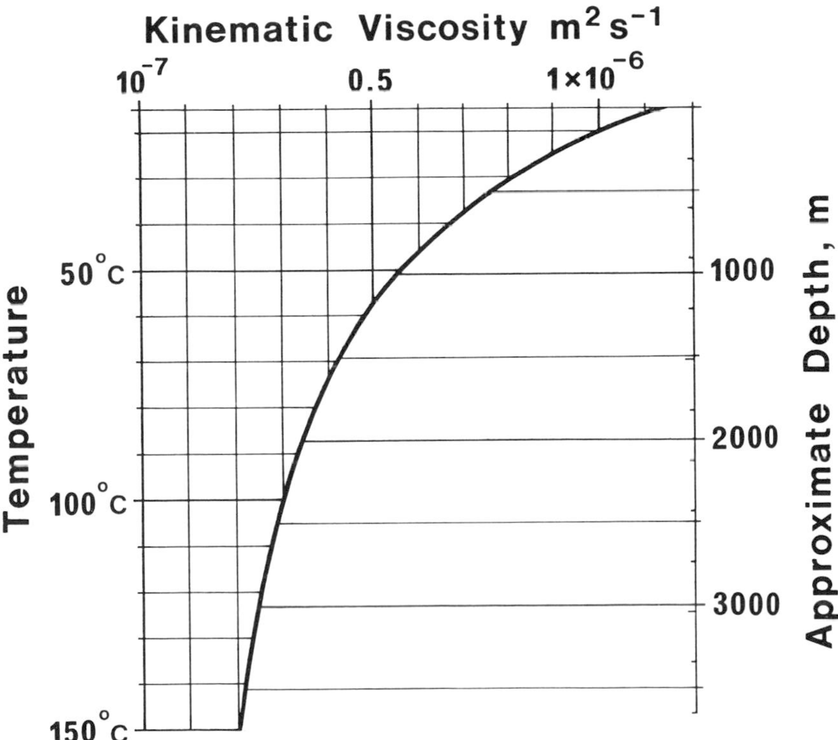

Fig. 3-11. Approximate kinematic viscosity of water under subsurface conditions with a geothermal gradient of 36°C/km and a normal hydrostatic pressure gradient (interpolated from data of Gray, 1957, p. 2-138, table 2n-1, and p. 2-209, table 2v-6).

significant error is likely if the average is taken. (If these sound like modern arguments, the reader is referred to King, 1899, pp. 81-82!)

These principles can be extended to multi-layer sequences, and the relative permeabilities of the layers deduced. The absolute values of the permeabilities, and of the specific discharge q, are much more difficult to obtain, and no attempt to do so will be made here.

Lateral migration of water through a sand may involve both changes of permeability and of the gross geometry of the sand. These will be considered in the next chapter after the introduction of flow nets; but it is worth noting here that the specific discharge will not be constant in a permeable bed of variable thickness because the cross-sectional area A changes. We therefore write equation (3.24)

$$Q = \mathrm{k}_1 \mathrm{A}_1 \frac{\rho}{\eta} g \frac{\Delta h_1}{l_1} = k_2 A_2 \frac{\rho}{\eta} g \frac{\Delta h_2}{l_2} \tag{3.26}$$

and A is commonly proportional to the thickness of the aquifer.

SELECTED BIBLIOGRAPHY

Archie, G.E., 1942. The electrical resistivity log as an aid in determining some reservoir characteristics. *Trans. American Inst. Mining Metallurgical Engineers* (Petroleum Division), 146: 54-62.

Bakhmeteff, B.A., and Feodoroff, N.V., 1937. Flow through granular media. *J. Applied Mechanics*, 4: A97-A104. (Also in *Trans. American Soc. Mechanical Engineers*, 59. See discussion by L.P. Hatch, *J. appl. Mech.*, 5: A86-A90, which is also in *Trans. Am. Soc. mech. Engrs*, 60).

Barrer, R.M., and Barrie, J.A., 1952. Sorption and surface diffusion in porous glass. *Proc. Royal Society*, A213: 250-265.

Bear, J., 1972. *Dynamics of fluids in porous media*. American Elsevier, New York, 764 pp.

Bett, K.E., and Cappi, J.B., 1965. Effect of pressure on the viscosity of water. *Nature*, 207 (4997): 620-621.

Blake, F.C., 1923. The resistance of packing to fluid flow. *Trans. American Inst. Chemical Engineers*, 14 (for 1922): 415-421.

Bredehoeft, J.D., and Hanshaw, B.B., 1968. On the maintenance of anomalous fluid pressures: I. Thick sedimentary sequences. *Bull. Geol. Soc. America*, 79: 1097-1106.

Bridgman, P.W., 1926. The effect of pressure on the viscosity of forty-three pure liquids. *Proc. American Academy Arts Sciences*, 61 (3): 57-99.

Brownell, L.E., and Katz, D.L., 1947. Flow of fluids through porous media. I. Single homogeneous fluids. *Chemical Engineering Progress*, 43 (10): 537-544. (Discussion: 544-548.)

Cornell, D., and Katz, D.L., 1953. Flow of gases through consolidated porous media. *Industrial and Engineering Chemistry*, 45 (10): 2145-2152.

Darcy, H., 1856. *Les fontaines publiques de la ville de Dijon*. V. Dalmont, Paris, 674 pp.

Debye, P., and Cleland, R.L., 1959. Flow of liquid hydrocarbons in porous *Vycor*. *J. Applied Physics*, 30 (6): 843-849.

Fair, G.M., and Hatch, L.P., 1933. Fundamental factors governing the stream-line flow of water through sand. *J. American Water Works Ass.*, 25 (11): 1551-1565.

Füchtbauer, H., 1967. Influence of different types of diagenesis on sandstone porosity. *Proc. 7th World Petroleum Congress*, 2: 353-370.

Gray, D.E., (Ed.), 1957. *American Institute of Physics Handbook* (1957). McGraw-Hill, New York.

Hansbo, S., 1960. Consolidation of clay with special reference to the influence of vertical sand drains. *Sweden, Statens Geotekniska Institut, Proc.*, 18: 1-160.

Happel, J., and Brenner, H., 1965. *Low Reynolds number hydrodynamics with special applications to particulate media*. Prentice-Hall, Englewood Cliffs, N.J., 553 pp.

Hubbert, M.K., 1940. Theory of ground-water motion. *J. Geology*, 48 (8): 785-944.

Hubbert, M.K., 1956. Darcy's law and the field equations of the flow of underground fluids. *Trans. American Inst. Mining Metallurgical Petroleum Engineers*, 207: 222-239. (Also in *J. Petroleum Technology*, 8.)

Hubbert, M.K., 1957. Darcy's law and the field equations of the flow of underground fluids. *Bulletin de l'Association Internationale d'Hydrologie Scientifique*, no. 5: 24-59.

Hubbert, M.K., 1969. *The theory of ground-water motion and related papers*. Hafner, New York and London, 310 pp.

Irmay, S., 1953. Saturated steady flow in non-homogeneous media and its applications to earth embankments, wells, drains. *Proc. 3rd International Conference Soil Mechanics* (Zürich), 2: 259-263.

Keller, G.V., 1966. Electrical properties of rocks and minerals. *In:* S.P. Clark (Ed.), Handbook of physical constants (revised edition). *Memoir Geol. Soc. America*, 97: 553-577.

Kennedy, G.C., and Holser, W.T., 1966. Pressure-volume-temperature and phase relations of water and carbon dioxide. *In*: S.P. Clark (Ed.), Handbook of physical constants (revised edition). *Memoir Geol. Soc. America*, 97: 371-384.

King, F.H., 1899. Principles and conditions of the movements of ground water. *Annual Report U.S. Geol. Surv.* (1897-98), 19 (2): 59-294.

Klinkenberg, L.J., 1942. The permeability of porous media to liquids and gases. *American Petroleum Inst. Drilling and Production Practice* 1941: 200-213.

Kozeny, J., 1927a. Über Grundwasserbewegung. *Wasserkraft und Wasserwirtschaft*, 22: 67-70, 86-88, 103-104, 120-122, 146-148.

Kozeny, J., 1927b. Über kapillare Leitung des Wassers im Boden (Aufstieg, Versickerung und Anwendung auf die Bewässerung). *Sitzungsberichte der Akademie der Wissenschaften in Wien*, Abt. IIa, 136: 271-306.

Lindquist, E., 1933. On the flow of water through porous soil. 1er *Congrès des Grands Barrages* (Stockholm, 1933), 5: 81-101.

Lutz, J.F., and Kemper, W.D., 1959. Intrinsic permeability of clay as affected by clay-water interaction. *Soil Science*, 88: 83-90.

Magara, K., 1971. Permeability considerations in generation of abnormal pressures. *J. Soc. Petroleum Engineers*, 11: 236-242.

Miller, R.J., and Low, P.F., 1963. Threshold gradient for water flow in clay systems. *Proc. Soil Soc. America*, 27: 605-609.

Nielsen, R.F., 1951. Permeability constancy range of a porous medium. *World Oil*, 132 (6): 188-192.

Nordberg, M.E., 1944. Properties of some *Vycor*-brand glasses. *J. American Ceramic Soc.*, 27: 299-305.

Norton, W.H., 1897. Artesian wells of Iowa. *Iowa Geol. Survey*, 6: 113-428.

Olsen, H.W., 1965. Deviations from Darcy's law in saturated clays. *Proc. Soil Science Soc. America*, 29: 135-140.

Olsen, H.W., 1966. Darcy's law in saturated kaolinite. *Water Resources Research*, 2: 287-296.

Owen, J.E., 1952. The resistivity of a fluid-filled porous body. *Trans. American Inst. Mining Metallurgical Engineers* (Petroleum Branch), 195: 169-174.

Plain, G.J., and Morrison, H.L., 1954. Critical Reynolds number and flow permeability. *American J. Physics*, 22: 143-146.

Rose, W.D., 1959. Calculations based on the Kozeny-Carman theory. *J. Geophysical Research*, 64 (1): 103-110.

Scheidegger, A.E., 1960. *The physics of flow through porous media* (revised edition). University of Toronto Press, Toronto, 313 pp.

Schneebeli, G., 1955. Expériences sur la limite de validité de la loi de Darcy et l'apparition de la turbulence dans un écoulement de filtration. *La Houille Blanche*, 10 (2): 141-149.

Schriever, W., 1930. Law of flow for the passage of a gas-free liquid through a spherical-grain sand. *Trans. American Inst. Mining Metallurgical Engineers* (PetroleumDivision), 86: 329-336.

Slichter, C.S., 1899. Theoretical investigation of the motion of ground waters. *Annual Report U.S. Geol. Surv.* (1897-98), 19 (2): 295-384.

Street, N., 1958. Tortuosity concepts. *Australian J. Chemistry*, 11 (4): 607-609.

Versluys, J., 1917. De beweging van het grondwater. *Water*, 1: 23-25, 44-46, 74-76, 95.

Versluys, J., 1930. The origin of artesian pressure. *Verhandelingen der Koninklijke Nederlandsche Akademie van Wetenschappen*, Afd. Natuurkunde, 33: 214-222.

Von Engelhardt, W., and Tunn, W., 1954. Über das Strömen von Flüssigkeiten durch Sandsteine. *Heidelberger Beiträge zur Mineralogie und Petrographie*, 4: 12-25.

Von Engelhardt, W., and Tunn, W.L.M., 1955. The flow of fluids through sandstone. *Illinois State Geol. Surv. Circular* 194 (17 pp.).

Weyer, K.E., 1978. Hydraulic forces in permeable media. *Mémoire du Bureau de Recherches Géologiques et Minières*, 91: 285-297.

Winsauer, W.O., Shearin, H.M., Masson, P.H., and Williams, M., 1952. Resistivity of brine-saturated sands in relation to pore geometry. *Bull. American Ass. Petroleum Geologists*, 36 (2): 253-277.

Zoback, M.D., and Byerlee, J.D., 1975. Permeability and effective stress. *Bull. American Ass. Petroleum Geologists*, 59 (1): 154-158.

4. THE AQUIFER AND FIELDS OF FLOW

It is one thing to make experiments to determine, as Darcy did, the amount of water that can be passed through a sand filter: it is quite another to apply these results to the geological materials of an aquifer. Consider an artesian aquifer, as in Figure 4-1, with water entering it in the intake area on high ground and leaving it by leakage or extraction where the ground is lower (to the left of the figure). Any well drilled into this aquifer will encounter water in it that will flow at the surface unless it is restrained by wellhead equipment. The pressure of the water and its density can be measured at the surface, and so the pressure head computed (if the density varies well to well, mean or unit density is taken). We can assume that the velocity head will be negligible, and the total head measured above our sea-level datum is

$$h = z + \frac{p}{\rho g} \tag{4.1}$$

and the fluid potential Φ in the position of the well is

$$\Phi = gh = gz + \frac{p}{\rho}. \tag{4.2}$$

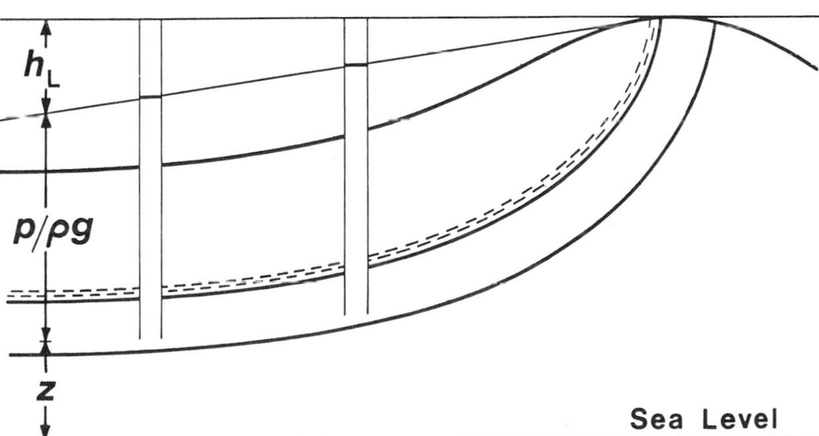

Fig. 4-1. Schematic section through leaking artesian basin, showing elevation head, pressure head, and head loss.

For each well the total head or fluid potential is easily calculated; and if there are enough wells, it will be possible to contour a surface that is the conceptual surface of total head of the aquifer and, lacking only the factor g, its fluid potential surface. This surface is known as the *potentiometric surface* (also, but less desirably, piezometric surface: it is not a surface that is a measure of pressure only). It is important for the reader to be absolutely clear as to what this surface is, and what it is not.

The potentiometric surface is a conceptual surface defined by the data of equations (4.1) or (4.2) in the aquifer. It is not the level to which the water would rise if it were free to do so: it is the level to which water would rise in a manometer inserted into the aquifer at any point. The shape of the surface is determined by the energy losses due to movement of the water in the aquifer: if the water is not moving, there is no energy loss and the potentiometric surface is horizontal. Contours drawn on the potentiometric surface are, of course, lines of equal potential that are known as *equipotential lines*. They are the intersection of an *equipotential surface* with the potentiometric surface. The equipotential surface only coincides with the potentiometric surface when the water is at rest: under all other conditions, they are inclined to each other.

We concluded from Darcy's experiments that water flows from positions of higher energy or potential to positions of lower energy or potential. Hence, where the potentiometric surface is not horizontal, the aquifer water from which it was derived is in motion, and it is *moving in the direction of maximum slope down the potentiometric surface*. Intuitively, we can accept that the direction of flow at a point is normal to the equipotential surface at that point, and so normal to the equipotential line at that point (it can also be proved analytically), so the macroscopic streamlines or *flowlines* can be drawn from the equipotential lines and the two form a *flow net* (Fig. 4-2). In general, flowlines converge where the liquid is accelerating, diverge where it is decelerating.

Consider two wells sited down a single flow line (but not producing), as in Figure 4-3. The water is flowing because the potentiometric surface is not horizontal: it is flowing from B to A because that is the direction of decreasing energy. The questions then arise, how much water is flowing, and how fast?

If we knew the hydraulic conductivity or intrinsic permeability of the aquifer we could at least answer the first question from Darcy's law:

$$q = \frac{Q}{A} = K\frac{h_B - h_A}{l} = k\frac{\rho}{\eta}\frac{\Delta\Phi}{l}. \tag{4.3}$$

(Note carefully that l is the macroscopic length through the aquifer, not the horizontal distance, so the slope of the potentiometric surface does not necessarily give the hydraulic or potential gradient.)

The practical problem here is that a measurement of permeability from a borehole core is, in the first place, a very small sample of the aquifer; and in the second place, it is likely to be mechanically disturbed, having been cut around by the bit and then

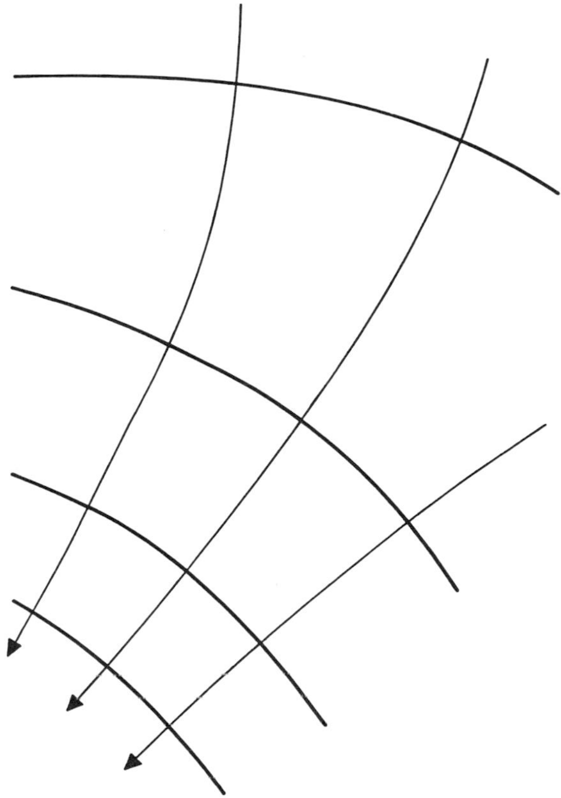

Fig. 4-2. Flow net formed by equipotential and flow lines.

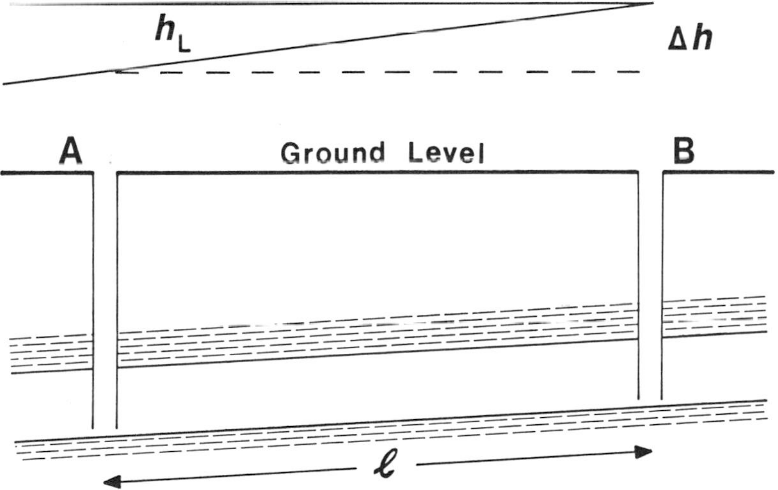

Fig. 4-3.

removed from its *in situ* temperature and pressure to atmospheric. So while its permeability can be measured with great precision, it is an illusory precision. (This is a common feature of subsurface geology that may beguile us in the belief that 'at last we have some hard data'.) Ideally, we wish to know the bulk permeability of the aquifer in the direction of flow, but must often be satisfied with estimates for lesser volumes – but volumes larger than the hand specimen. The best estimates available with present technology are from data obtained from pumping wells. Before considering this, it is necessary to look at the field of flow within the aquifer more closely.

Fields of flow

Any physical quantity that can be mapped in 3-dimensional space can be said to have a *field* of that quantity. Thus a body of water can be said to have a temperature field, a pressure field, and, if the water is moving, a velocity field. Since temperature and pressure are *scalar* quantities (for any position they are sufficiently defined by a single number and a scale of measurement) and velocity is a *vector* quantity (for any position, number, scale, and direction are required for its definition) the body of water is more precisely said to have scalar fields of temperature and pressure, and a vector field of velocity. We talk too of gravity and stress fields (the latter being important in matters we shall come to in later chapters). Gravity is a vector quantity that, for our purposes, can be regarded as a field in which the magnitude and direction of the force remains constant (leaving to geophysicists the small variations that are significant for them). Stress is a *tensor* quantity, requiring six quantities for its definition (tensors are to vectors what vectors are to scalars). For the moment, we shall confine ourselves to scalar and vector fields in the context of water flow through porous rocks.

For every scalar field there is an associated vector field of the rate of change of the scalar quantity in 3-dimensional space. For example, in a body of water there are surfaces in the pressure field that pass continuously through points of equal pressure. These surfaces of equal pressure (isobaric surfaces) in static water are horizontal: but there is a vector field of pressure gradient in which the field lines are everywhere normal to the isobaric surfaces. Likewise, there is a vector field of temperature gradient in which the field lines are normal to the isothermal surfaces.

Potentials are scalar quantities: the potential at a point has magnitude but no direction. Hence the fluid potential field in a body of water is a scalar field, with which is associated a vector field of potential gradient. This field is of fundamental importance in the analysis of ground-water movement because the fluid potential gradient is, as we have seen, the source of the 'driving force' of the water through porous rock.

There are four fields in a ground-water aquifer that are of particular importance: the pressure field, the velocity field, the potential field – and all geological processes take place in the gravity field. We shall first consider the pressure field and the gravity field.

When a body of liquid is at rest, the surfaces of equal pressure are all horizontal and the vector field of pressure gradient has perpendicular field lines that are sometimes said to represent vectors equal in magnitude but opposite in direction to the force of gravity. This is wrong because the dimensions of pressure gradient are $[ML^{-1}T^{-2}][L^{-1}] = [ML^{-2}T^{-2}]$, which are the dimensions of weight density, ρg, as we would expect, while the dimensions of force (of gravity) are $[MLT^{-2}]$. We must compare only quantities that are dimensionally alike. The force of gravity gives to unit mass an acceleration g. What, then, is the pressure as a force on unit mass? Pressure is a force per unit of area on which it acts – in this case, a horizontal area. The pressure gradient is the change of pressure along a line normal to the isobaric surfaces – in this case, a vertical line. Hence the net force acting on a unit cube of the liquid (Fig. 4-4) is the difference between the pressures on the two horizontal faces of the unit cube, $(p_2 - p_1)$. The mass of this unit volume of the liquid is ρ. Hence the net force acting on unit mass of the liquid is

$$F_p = \frac{1}{\rho}\frac{p_2 - p_1}{d_2 - d_1} = \frac{1}{\rho} \text{ grad } p, \qquad (4.4)$$

where $(p_2 - p_1)/(d_2 - d_1)$ is the pressure gradient, which we can conveniently abbreviate by using the vector notation, grad p. This force is dimensionally identical

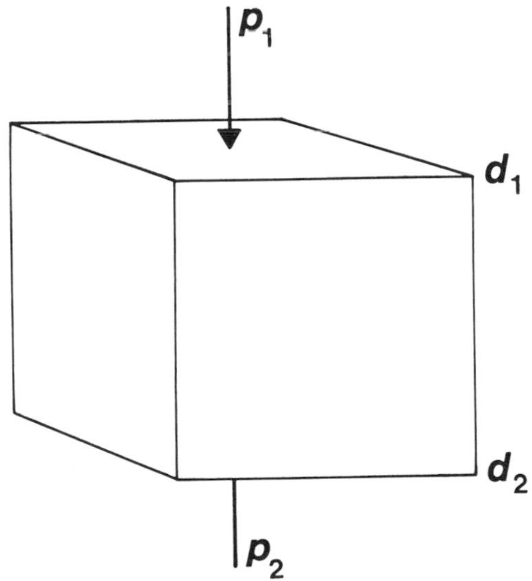

Fig. 4-4.

to the force of gravity acting on unit mass (that is, an acceleration, LT^{-2}) and since grad p is a vector of magnitude ρg in a fluid at rest,

$$\frac{1}{\rho}\,\text{grad}\, p = \mathbf{g} \qquad (4.5)$$

and we can say that the vector field of pressure gradient in terms of force per unit of mass in a body of water at rest is equal in magnitude and opposite in direction to the vector field of gravity. The resultant force on the body of water is zero, and it is therefore at rest (Fig. 4-5).

By continuing this line of argument we can say that the isobaric surfaces in a body of water in motion are not horizontal, and the vector field of grad p is not vertical (Fig. 4-6). The resultant force acts on the water to give it an acceleration that is reduced to a velocity by the frictional resistive forces. But velocity of liquids in porous solids, as we have seen, is difficult to deal with. Let us therefore look at the fluid potential field.

The fluid potential Φ is the scalar quantity of energy possessed by unit mass of the liquid in a particular position. The associated vector quantity, grad Φ, is the fluid potential gradient and is the quantity that determines the direction of flow and, with the coefficient of permeability or hydraulic conductivity K, the specific discharge of water (volume across unit area in unit time). Hubbert (1940, pp. 794-802) showed that the fluid potential is the sum of the work done against gravity and pressure when transposing unit mass by a frictionless process from an arbitrary state and place of reference to the particular state and place of interest:

$$\Phi = gz + \frac{p - p_0}{\rho}. \qquad [L^2T^{-2}] \qquad (4.6)$$

We have just considered the second term on the right of equation (4.6); we now look at the first term on the right. In Chapter 2 (p. 30) we took gz to be the potential mechanical energy of unit mass of water, which we shall now call the *gravity potential*

$$U = gz. \qquad [L^2T^{-2}] \qquad (4.7)$$

This scalar field has its associated vector field of gradient of gravity potential with dimensions $[LT^{-2}]$. Clearly, the surfaces of equal gravity potential are horizontal, and the field lines of gravity potential gradient are vertical and equal to **g**. If the body of water is static, the surfaces of equal pressure are horizontal, with vertical field lines of pressure gradient in terms of force per unit mass; and in this special case, the vector $1/\rho$ grad p is equal in magnitude and opposite in direction to the gravity potential gradient vector, grad U. The resultant force is zero. If the body of water is moving, however, the vector quantity of force per unit of mass due to pressure is not vertical, nor is it necessarily equal to grad U; and the resultant is the vector quantity of *fluid potential gradient* (Fig. 4-7).

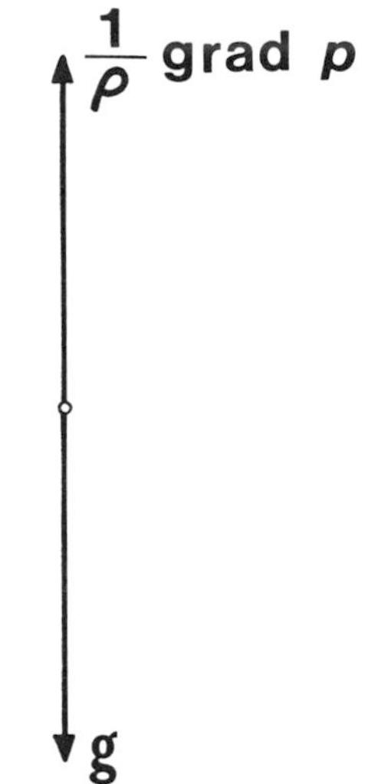

Fig. 4-5.

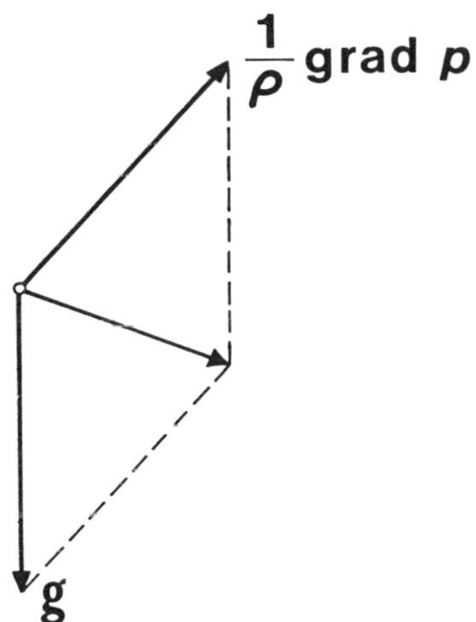

Fig. 4-6.

We thus have three superimposed fields of which the field of fluid potential gradient is the resultant of the other two. Surfaces of constant pressure, gravity potential, and fluid potential are normal to their gradient vectors (Fig. 4-8); and, in theory at least, they are mappable from data acquired from boreholes. Ground water moves in the direction of the fluid potential gradient vector, from positions of higher fluid potential to positions of lower fluid potential.

Reverting now to the artesian aquifer with which we started this chapter, we can infer the fields of flow within the aquifer from the measurements of depths and pressures in boreholes, and the computed potentiometric surface (Fig. 4-9). Because the water is confined to its aquifer, its macroscopic flow lines are restricted by the

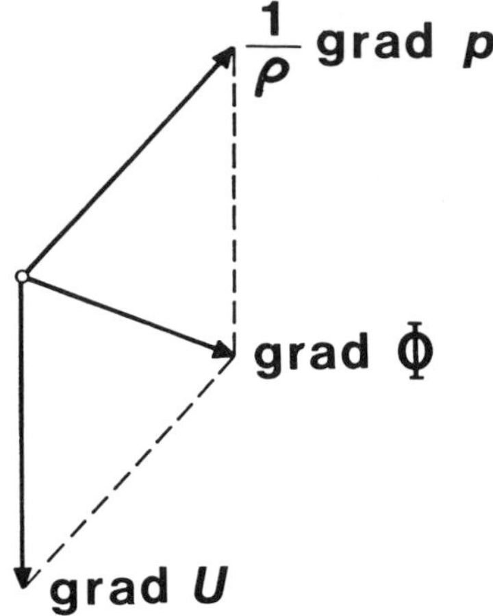

Fig. 4-7.

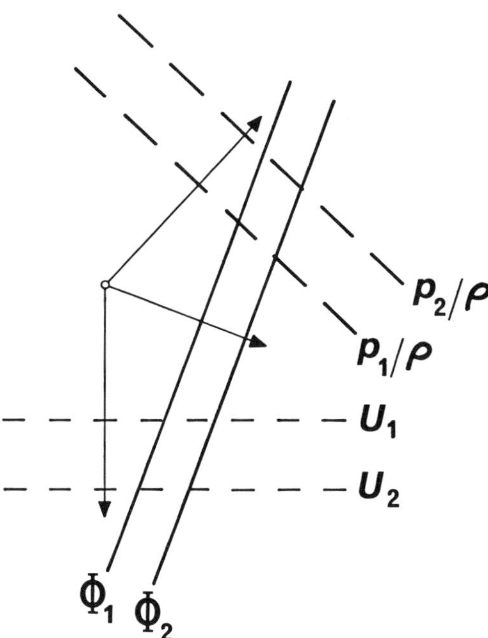

Fig. 4-8. Surfaces of constant pressure, gravity potential, and fluid potential are normal to their gradient vectors.

upper and lower boundaries of the aquifer, and the equipotential surfaces are normal to this flow. The surfaces of equal pressure are not horizontal, but inclined in the direction of motion by an amount that is determined, in effect, by the fluid potential gradient.

It will be remembered that grad Φ is a vector, which has both direction and magnitude. The direction conforms to the aquifer, as it must: the magnitude may be

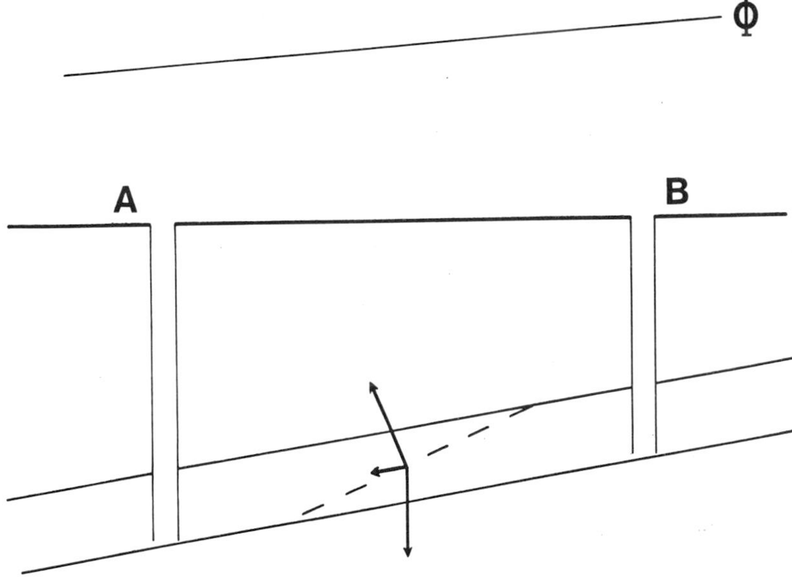

Fig. 4-9. Gradient vectors of pressure, gravity potential, and fluid potential in a confined aquifer (exaggerated).

large or small. In artesian basins, the fluid potential is usually relatively large, but the fluid potential gradient is small unless there is significant leakage or production from the aquifer.

Now, the volume of the aquifer enclosed by two flow lines projected through the aquifer (Fig. 4-10) may be considered as a stream tube, or flow tube, enclosed by an impermeable surface (because there can be no flow across flow lines). It follows that the mass or volume discharge in steady flow across any transverse area of the flow tube per unit of time is constant. Thus, from Darcy's law,

$$\begin{aligned} Q &= K_1 A_1 \Delta h_1 / l_1 \\ &= K_2 A_2 \Delta h_2 / l_2. \end{aligned} \tag{4.8}$$

If we assume that the hydraulic gradient, $\Delta h/l$, is approximated by the contour spacing (as it is in all gently-dipping aquifers) we can see that

1) if A remains constant, K is greatest when the hydraulic gradient is least;
2) if A decreases downstream and K remains constant, the hydraulic gradient increases; and
3) if the hydraulic gradient remains constant, K increases as A decreases.

Thus a map of the potentiometric surface of an aquifer allows us to infer the directions of water movement within the aquifer, and to make certain general deductions about the changes in aquifer properties. It does more: a closed area of high potential can only exist where there is a source, and water is being added to the

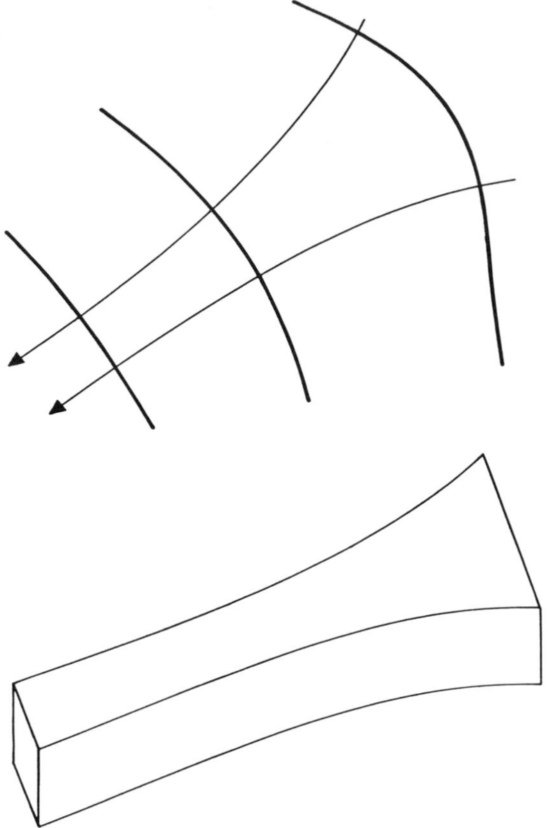

Fig. 4-10. Stream tube in confined aquifer enclosed by two flow lines projected through the aquifer.

aquifer. And a closed area of low potential can only exist where there is a sink, and water is being taken from the aquifer. The latter is of present interest to us because it happens around a producing well, and the form of the sink (known as the cone of depression) enables us to obtain a measure of permeability over a larger area than the hand specimen.

PRODUCING WATER WELL

Consider a well that penetrates a horizontal isotropic confined aquifer completely, the water in the aquifer being static (Fig. 4-11). Before the well produces, the level of the water in it coincides with the potentiometric surface of the water in the aquifer. Isobaric surfaces are horizontal; the water is everywhere within the aquifer at the same potential, so there is one horizontal equipotential surface: the water does not flow. When water is produced from the well, the well becomes a sink, and water

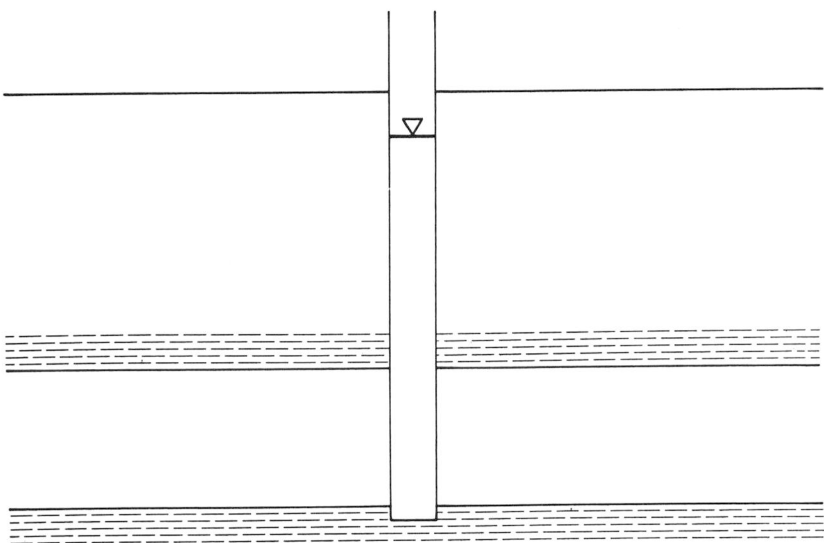

Fig. 4-11.

flows radially through the aquifer to the well down a potential gradient that is created by the reduced potential in the well. The volume of water, Q, produced in unit time (when steady flow has been achieved, which may take some time) is produced into the borehole across an area

$$A_w = 2\pi r_w t, \tag{4.9}$$

where r_w is the radius of the borehole, and t the thickness of the aquifer. The same volume of water is produced in unit time across all concentric areas of greater radius, r, so that

$$A = 2\pi r t. \tag{4.9a}$$

Letting q_w represent the specific discharge, Q/A_w, into the well; and q, the specific discharge across the concentric area A at radius r, then

$$q_w \times 2\pi r_w t = q \times 2\pi r t$$

and

$$q = q_w r_w / r. \tag{4.10}$$

From this we see that the specific discharge, which is a notional velocity, varies inversely with distance from the well. From Darcy's law

$$q = K\frac{\Delta h}{\Delta r} = q_w r_w / r \tag{4.11}$$

from which the hydraulic gradient is seen to be

$$\frac{\Delta h}{\Delta r} = \frac{q_w}{K}\frac{r_w}{r}. \tag{4.11a}$$

Indicating the total head at the wall of the borehole by h_w, the total head at radius r is given by

$$h_r = h_w + \frac{q_w r_w}{K}\int_{r_w}^{r}\frac{1}{r}\,dr$$

$$= h_w + \frac{q_w r_w}{K}\ln\frac{r}{r_w}. \tag{4.12}$$

Since $q_w = Q/2\pi r_w t$, we may substitute $Q/2\pi t$ for $q_w r_w$ in equation (4.12):

$$h_r = h_w + \frac{Q}{2\pi t K}\ln\frac{r}{r_w} \tag{4.12a}$$

and the total head is seen to increase as the natural logarithm of the distance from the borehole, approaching the total head of the static aquifer (prior to production from the well).

Equation (4.12a) indicates that h increases indefinitely as r increases, that is, steady flow is theoretically impossible in a finite aquifer.

More generally,

$$h_2 - h_1 = \frac{Q}{2\pi t K}\ln\frac{r_2}{r_1}, \tag{4.13}$$

where the suffix 2 represents a station further from the well than the station represented by the suffix 1 (Fig. 4-12), which may be at the producing well. The boundary conditions for these equations are that $h_2 = h$ at $r = r_i$, where r_i is the 'radius of influence'. Equation (4.13) is due to Thiem (1906) and bears his name. The product tK, with dimensions $[L^2T^{-1}]$, is known as the *transmissivity* of the aquifer.

The hydraulic conductivity is therefore given by

$$K = \frac{Q}{2\pi t(h_2 - h_1)}\ln\frac{r_1}{r_2}. \tag{4.13a}$$

From this it can be seen that if we can determine the elevation of the potentiometric surface in two positions when the well is producing steadily at rate Q, then the value of the hydraulic conductivity can be determined. The elevation of the potentiometric surface can be measured in what are called 'witness wells' (boreholes of small diameter drilled specially for this purpose).

Equation (4.13a) is strictly valid only for steady flow, and this is theoretically impossible in a finite aquifer. But it shows that the larger the value of K, the shallower the cone of depression and the larger the radius of influence of the producing well. Witness wells should be drilled well within the radius of influence, in the pronounced part of the cone. The producing well itself may be used to determine

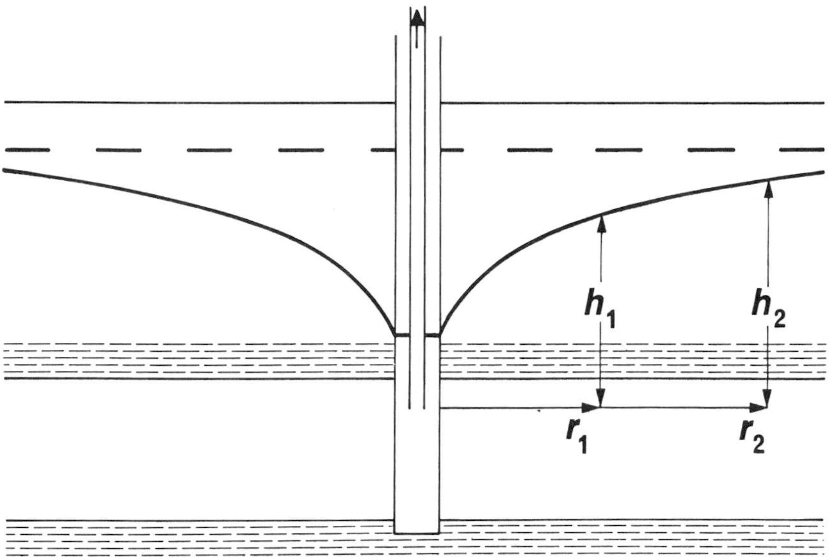

Fig. 4-12.

h_1: but while this practice leads to a risk of error due to possible non-Darcy behaviour near the borehole, it should never be forgotten that it is better than no data at all. The producing well need not be produced at the maximum rate for the purposes of determining the hydraulic conductivity, but it must be at a rate high enough to generate a measurable cone of depression or *drawdown* in the well. Steady flow may be assumed when the water level (or pressure) in the witness wells stabilize.

If the water in the aquifer is not static, that is, the potentiometric surface is inclined from the horizontal, then the symmetry of the cone of depression is lost when the borehole is put on production. But potentials and heads are scalar quantities, so the observed potentials or heads are the result of the superposition of one field (due to pumping) on another:

$$\left.\begin{aligned} \Phi_{\text{total}} &= \Phi_{\text{aquifer}} + \Phi_{\text{well}}\,. \\ h_{\text{total}} &= h_{\text{aquifer}} + h_{\text{well}}\,. \end{aligned}\right\} \tag{4.14}$$

The potential or head due to the well will be negative when water is being extracted (positive, if being injected).

If this refinement is to be pursued, and it need only be pursued when the inclination of the aquifer's potentiometric surface is considerable, at least two witness wells will be required, and they must not be in a straight line with the well. If such a configuration as in Figure 4-13 is used, we can aquire valuable information on the aquifer. In the first place, the static water levels in the three wells (static, in this context, refers to the levels when the well is not producing, or the aquifer has

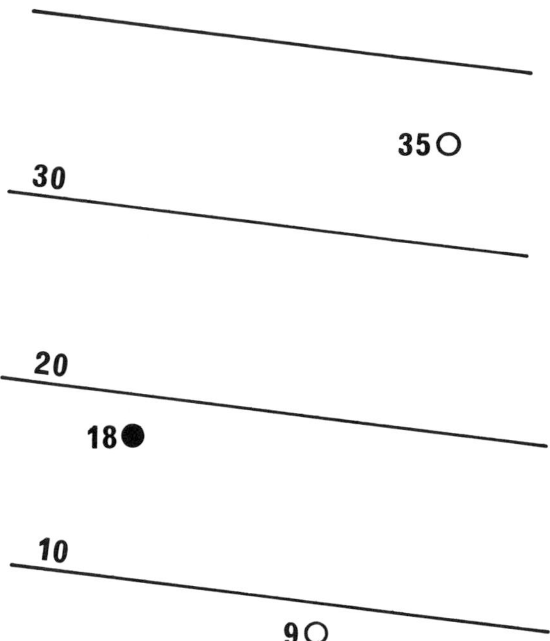

Fig. 4-13.

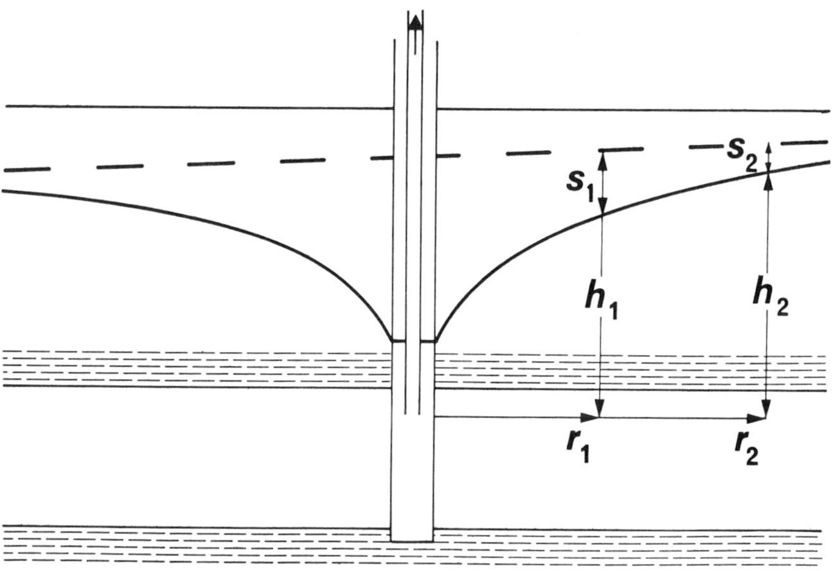

Fig. 4-14a. Potentiometric surface in profile across well producing from a confined aquifer in which the regional flow is from right to left.

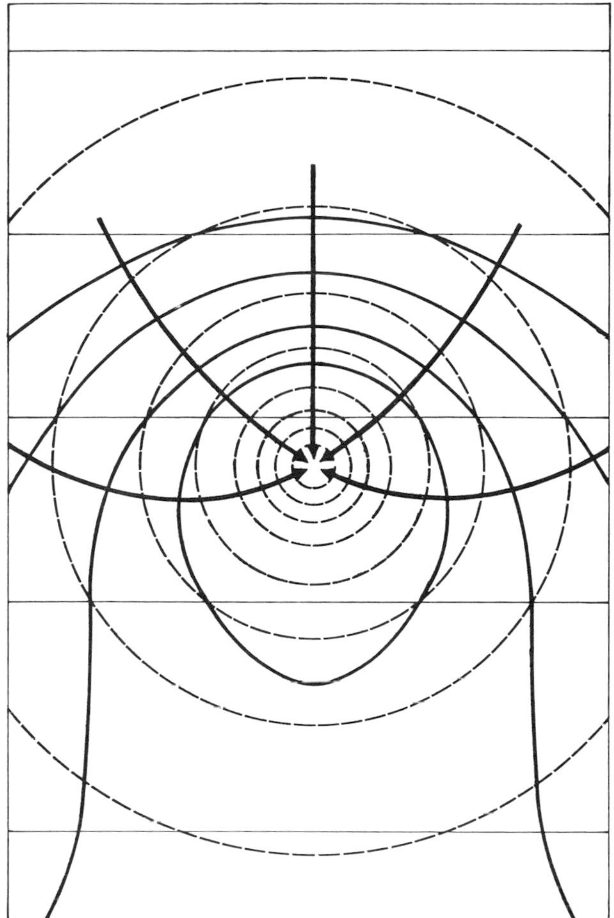

Fig. 4-14b. Flow net obtained by superimposing the effect of the well on the aquifer.

stabilized after production) define the slope of the potentiometric surface, and hence the direction of water movement in the aquifer.

When steady production at rate Q has been attained, the cone of depression that results is superimposed on the potentiometric surface of the aquifer. There are, of course, sophisticated methods of solving these problems, but a simple approach may still be useful. The drawdown (s) is the difference between the static and pumping levels in the wells (Equation (4.13a) could have been written in terms of drawdown rather than head). Hence, an approximation to the hydraulic conductivity can be obtained by substituting $(s_1 - s_2)$ for $(h_2 - h_1)$ in equation (4.13a); and from equation (4.13), the radii of equal intervals of drawdown can be estimated. The equipotential lines of the combined field are obtained from the algebraic sum of the two components, as in Figure 4-14. The flow net can be completed by sketching in the flow lines normal to the equipotential lines.

There is a relatively narrow path to follow in these matters: on the one hand, there is the danger of drawing inferences that are not justified by the data, and, on the other hand, the danger of being so scared by the lack of precision that no assessment is made. For example, if the hydraulic conductivity can be estimated from the drawdown data, this also allows us to estimate the natural flow through the aquifer in the vicinity of the well when it is not on production. This estimate may well be inaccurate; but if reasonable care is taken, it is surely better than no estimate.

Unconfined aquifers

We have considered confined aquifers so far because they are more easily treated quantitatively. Unconfined aquifers are more difficult because the water table, or free water surface, is the potentiometric surface (Fig. 4-15). Not only is the flow then three-dimensional near the well, but also the area of concentric cylinders across which the water flows decreases towards the well (c.f., t, the aquifer thickness, assumed constant in the confined aquifer).

The equation for unconfined flow, analogous to equation (4.13), is

$$h_2^2 - h_1^2 = \frac{Q}{\pi K} \ln \frac{r_2}{r_1} \tag{4.15}$$

but this equation lacks the rigour of equation (4.13) because it depends on assumptions known as the *Dupuit assumptions*, which are: that the flow is uniform and horizontal, and that the flow rate is proportional to the tangent of the free surface slope, $\Delta h/\Delta r$, not the sine. These assumptions are clearly reasonable for flow at some

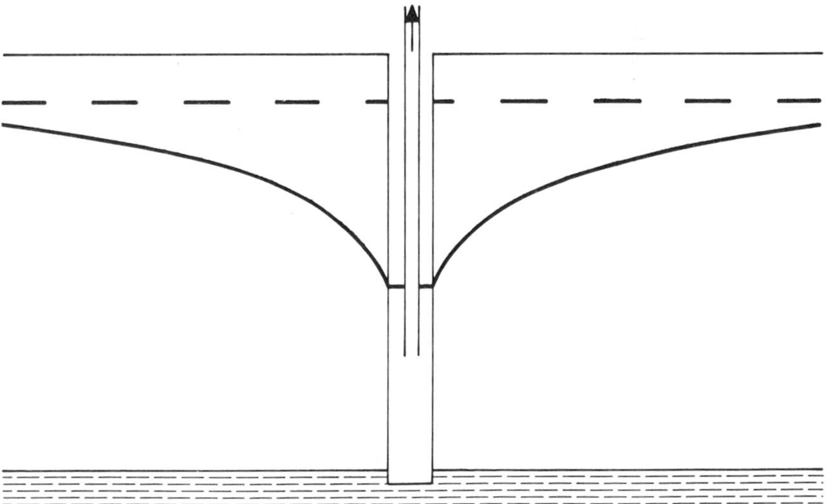

Fig. 4-15. Well penetrating unconfined aquifer: production induces a cone of depression on the water-table.

distance from the well, where the slope of the free water surface is small, but they become quite unacceptable near the well, where there is a considerable vertical component of flow.

Equation (4.15) is derived as follows:
Assume a horizontal isotropic aquifer with an impermeable base and static water to be penetrated completely by a well producing steadily at rate Q. With the assumption of horizontal flow, this rate Q flows across any concentric cylindrical surface below the free water level: the area of such a surface is is $2\pi rh$ where h is the elevation of the free water surface above the base of the aquifer. Darcy's law, with the Dupuit assumptions, is then written

$$Q = KA\frac{dh}{dr}$$

$$= 2\pi Krh\frac{dh}{dr}. \qquad (4.16)$$

Integration of this equation with respect to h and to r, for the limits $h = h_1$ when $r = r_1$ and $h = h_2$ when $r = r_2$, gives us

$$Q = \pi K\frac{h_2^2 - h_1^2}{\ln(r_2/r_1)}$$

which can be re-arranged in the form of equation (4.15).

Provided the slope dh/dr is not large, an estimate of the hydraulic conductivity can be obtained from a producing well and *two* witness wells. If r_1 is taken as the borehole radius, r_w, or r_1 is small (witness well near the water well), such estimates can be misleading because the Dupuit assumptions are no longer valid.

It will occur to the reader that when a well ceases production from either a confined or an unconfined aquifer, the rate at which cone of depression is eliminated, and the rate of drawn-down water levels in the wells returning to normal, will also be a measure of the hydraulic parameters of the aquifer (Horner, 1951). Likewise, if water is injected into the aquifer, there will be a cone of *impression* above the original free water surface; and its shape will depend on the hydraulic parameters of an unconfined aquifer (for a confined aquifer, the cone of impression is superimposed on the potentiometric surface).

It is not our purpose here to pursue these matters to the point that the reader will be equipped to perform the duties of a ground-water hydrologist; for that, he must consult such standard texts as Todd (1959), De Wiest (1965), and Davis and DeWiest (1966). Our aim is an understanding of the phenomena.

NATURAL SINKS

The phenomena of drawdown and discharge associated with water wells producing from unconfined aquifers have their natural counterpart in the flow of streams and rivers that penetrate an unconfined aquifer (Fig. 4-16). However, it is in the nature of an unconfined aquifer that it is subject to recharge from rainfall: so the conditions change with time, or from season to season.

To understand these processes, we shall look at water flow into an open channel that partly penetrates an unconfined aquifer. Consider the channel in Figure 4-16 to be partly penetrating an isotropic uniform unconfined aquifer of great thickness relative to the depth of the incision; and assume that the aquifer is uniformly recharged by rainfall over the area so that the water-table is maintained at a constant level in each position. The aquifer has a source, therefore, and the channel is a sink. Water is flowing through the aquifer from the ground-water divide to the channel, and our purpose is to sketch in the equipotential lines and the flow lines.

At equal elevations above the channel, on each side, an equipotential surface intersects the water table (Fig. 4-16). Since the water in the channel is at the lowest potential in the section, an equipotential surface close to the channel must pass beneath it, and flow will be radial to the channel. At the divide, the direction of flow changes from flow to the channel to flow to the adjacent channel: here the equipotential surfaces are horizontal, with flow vertically downwards. In the region of the divide, therefore, the equipotential surfaces are concave upwards.

We therefore infer that flow takes place in all directions from vertically downward at the water-table divide to vertically upward under the channel, and, following King (1899, p. 99, fig. 14) and Hubbert (1940, pp. 166-170) sketch the flow net as in Figure 4-17. On account of the rainfall recharge, assumed uniform over the area, the flow lines near the water table are steeper than the water-table slope.

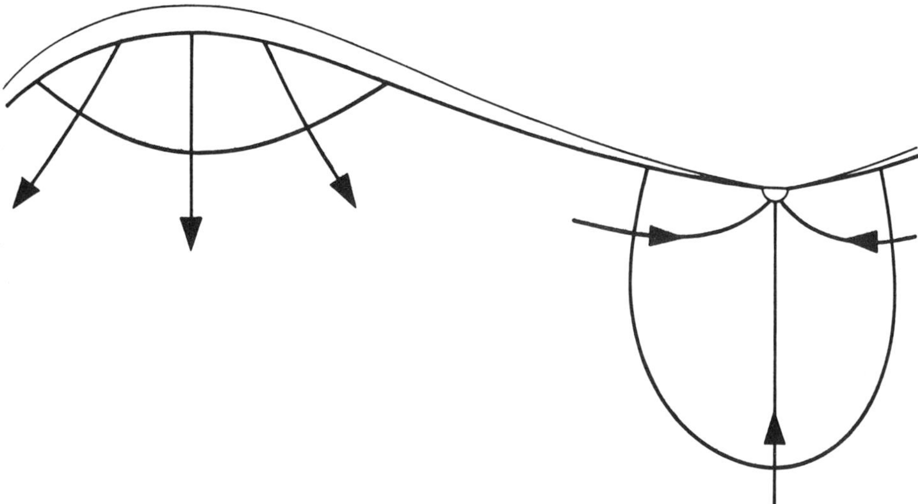

Fig. 4-16. Schematic section showing a stream penetrating an unconfined aquifer.

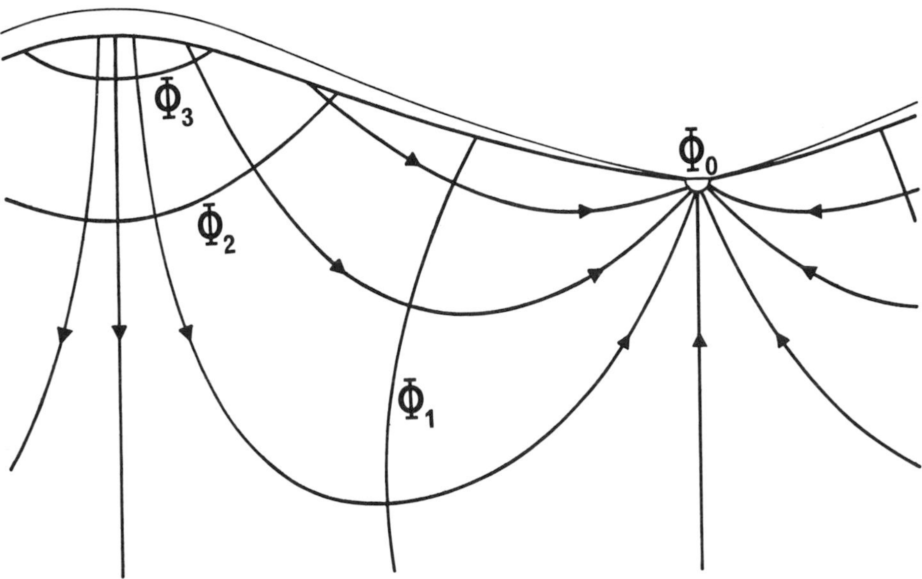

Fig. 4-17. Schematic flow-net in section for Figure 4-16.

A difficulty with this presentation is that one wonders how the equipotential surface lines between Φ_1 and Φ_2 should be drawn, because these have curvature in opposite senses, with Φ_1 never passing under the divide and Φ_2 never passing under the channel in this line of section. This difficulty can be understood to some extent by considering that the mean potential lies somewhere between Φ_1 and Φ_2, and the divide disturbs this in one direction, the sink in the other.

In plan (Fig. 4-18) we can also sketch the flow-net and see that equipotential surface pass under both divide and channel, and that in general, the flow lines follow the topographical slopes. This was clearly understood by Latham (1878), who wrote '...water standing at such a declivity is clear evidence of movement' and 'The greatest elevation of the subterranean water is usually found under the highest lands, and the least elevation under the lands having the lowest level. The flow of water laterally is from the hills to the valleys, and longitudinally down the valley-lines; therefore, as a general rule, the flow of subterranean water conforms to the surface-falls of the country' (Latham, 1878, p. 207).

In nature, of course, recharge is not uniform, nor are aquifers usually of very great thickness, nor are they isotropic. Nevertheless, the corollary of our first fundamental proposition (p. 5) is reinforced: when the free surface of a liquid is not horizontal, that liquid is in motion. And we infer motion *throughout the aquifer*, not just near the water table.

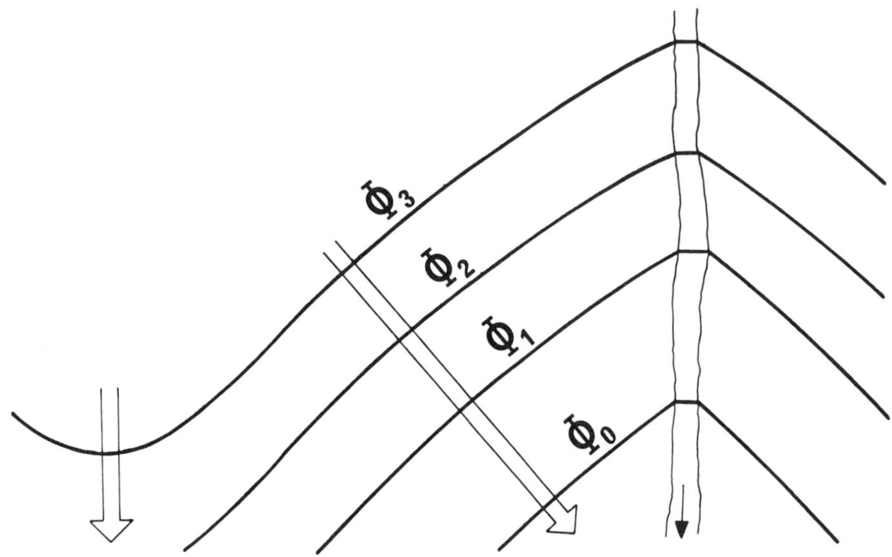

Fig. 4-18. Schematic flow-net in plan for Figure 4-16.

SELECTED BIBLIOGRAPHY

Davis, S.N., and DeWiest, R.J.M., 1966. *Hydrogeology*. John Wiley & Sons, New York, London, and Sydney, 463 pp.

De Wiest, R.J.M., 1965. *Geohydrology*. John Wiley & Sons, New York, London, and Sydney, 366 pp.

Heath, R.C., and Trainer, F.W., 1968. *Introduction to ground-water hydrology*. John Wiley & Sons, New York, London, and Sydney, 284 pp.

Horner, D.R., 1951. Pressure build-up in wells. *Proc. 3rd World Petroleum Congress* (The Hague, 1951), Section 2: 503-521.

Hubbert, M.K., 1940. The theory of ground-water motion. *J. Geology*, 48 (8): 785-944.

King, F.H., 1899. Principles and conditions of the movements of ground water. *Annual Report U.S. Geol. Surv.* (1897-98), 19 (2): 59-294.

Latham, B., 1878. Indications of the movement of subterranean water in the Chalk Formation. *Report British Ass. Advancement of Science* (47th Meeting, Plymouth, 1877): 207-209.

Meinzer, O.E., 1923. Outline of ground-water hydrology with definitions. *U.S. Geol. Surv. Water-Supply Paper* 494 (71 pp.). (*Read definitions critically.*)

Norton, W.H., 1897. Artesian wells of Iowa. *Iowa Geol. Survey*, 6: 113-428.

Thiem, G., 1906. *Hydrologische Methoden*. (Inaugural dissertation, Stuttgart). Gebhardt, Leipzig, 56 pp.

Todd, D.K., 1959. *Ground water hydrology*. John Wiley & Sons, New York and London, 336 pp.

5. AQUIFERS: SPRINGS, RIVERS, AND MAN-MADE DRAINAGE

Let us remind ourselves at the outset that rivers are not merely channels that carry rain-water runoff to the sea. Perennial rivers are fed by ground water, streams, and occasionally by rain-water runoff during and after storms: perennial streams are fed by ground water, smaller streams, and occasional runoff: the source is a spring or a zone of seepage, and is the highest intersection of the ground surface with the water-table. Common causes of springs are illustrated in Figure 5-1. When there is a depression in the ground that penetrates the water-table, there is a lake: but a lake may also be fed by a stream, and it may overflow as a stream. In these matters we are nearly always concerned with unconfined aquifers, which are recharged by rainfall that percolates downwards through the soil.

The flow of intermittent rivers may be due largely to runoff from storms (as desert *wadis*) or to the seasonal raising of the water table due to the wet season (as in many

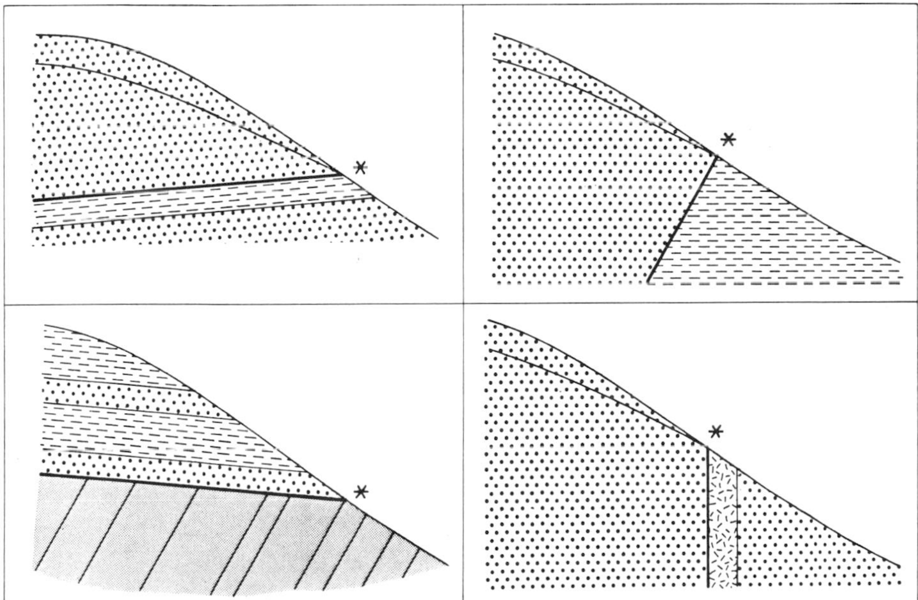

Fig. 5-1. Common causes of springs, in section.

Australian rivers). Also, reaches of a river may be intermittent where it passes over the outcrop of a porous and permeable formation in which the water table may from time to time lie well below the bed of the river (thick limestones are commonly the cause of this).

So the activity of springs and the flow of rivers and streams is related to rainfall and the resulting variations in the level of the water table.

These observations are not new, of course. The ancients thought that springs and rivers were fed by subterranean reservoirs (it would be arrogant of us to suppose that they all misunderstood the nature of these reservoirs), but the first scientific observations were carried out by Pierre Perrault in the upper Seine river basin in France between 1668 and 1670, and published in his *De l'origine des fontaines* in 1674. He measured the rainfall and estimated the river flow, and found that the quantity of water falling as rain was about six times the quantity carried by the river. These measurements dispelled any belief that rainfall was inadequate to account for the flow of rivers (see Perrault, 1967, for reference to an English translation of Perrault, 1674; but Perrault cannot be credited with *understanding* springs because throughout his book he insists that rain-water cannot penetrate deeply into the soil: he thought the waters in wells came *from* rivers – that if there were no rivers, there would be no water for the wells).

A little later, Edmé Mariotte verified Perrault's results in Paris (his work being published posthumously in 1686, two years after he died). He estimated the annual rainfall in the Seine catchment area above Paris to be about 7.14×10^{11} cubic *piés* (a little larger than an English foot). The river below Pont-Royal was 400 piés wide, and five deep, flowing at an average speed of about 150 piés/minute (250 when in flood); 'but,' he wrote, 'because the water at the bottom does not move as fast as that in the middle, nor the middle as fast as that at the surface ... one can take as the mean speed 100 piés in a minute'. Rainfall exceeded the river flow by a factor of nearly seven (Mariotte, 1717, pp. 338-339).

Mariotte also observed that the infiltration of water into a cellar at the Paris Observatory varied with the rainfall. He wrote (op. cit., p. 336) 'The summer of 1681 was very dry in France, so much so that most wells and springs in many areas dried up; and although the end of October and the beginning of November were rather cold, the waters continued to diminish, which they would not have done if water had been formed by vapours rising from below ground and condensing from the cold at the surface of the Earth. There is a hollow in the cellars of the Observatory in which there had been water continuously from 1668 to 1681: but the drought of 1681 dried it up entirely, and there was no longer a single drop in it in February 1682 although there had been much rain over several days at the beginning of this month; and in spite of the following summer having been very rainy, the water did not return in the month of September, nor even during the two following years'.

About the same time, the British Astronomer Edmond Halley determined the rate of evaporation from a pan of water salted to the salinity of the sea – and extrapolated his results to the Mediterranean Sea. The Mediterranean is, as its name suggests, a sea that is almost enclosed by land. Into it flow several large rivers: from the north, the Ebro, Rhône, and Po flow into the Mediterranean itself, and the Danube, Dniester, Dnieper, and Don rivers flow into the Black Sea, the only outlet from which is through the Bosphorus into the Mediterranean. From the south flows the Nile. Nor is the Mediterranean free from rain. The only connexion with an ocean is to the Atlantic through the Straits of Gibraltar, where the surface current also flows into the Mediterranean. Halley (1687) extrapolated his experimental results, made some gross approximations to the volumes of water contributed by the rivers, and concluded that evaporation was entirely adequate to remove the waters contributed by rivers and rainfall. He estimated that the contribution of the rivers was only one third of the amount removed by evaporation. This proportion is the same as that arrived at in modern times (see G[orgy] and S[alah], 1974, p. 856, Table 1). A few years later, Halley (1691) gave a brief but accurate 'account of the circulation of the watry Vapours of the Sea, and of the cause of Springs'.

Thus the essential ingredients of what is now known as the *hydrological cycle* (Fig. 5-2) were established more than two centuries ago. We shall concentrate on that part of the cycle that lies in and on the land, and the inter-relationships between aquifers, rivers, and rainfall, and some cases of human interference in these that serve to illustrate the inter-relationships.

It has no doubt been observed ever since wells were constructed that the water-level in wells fluctuates; but precise observations had to await the development of instruments capable of measuring these fluctuations with time. Such instruments enabled King (1892) and Veatch (1906) to publish accounts of detailed observations concerning the fluctuations of water-levels in wells – fluctuations mainly over periods of a few days, measured with great precision. These are still of interest (indeed, we mentioned King's observations on changes of water-level due to loading by a railway train at the end of Chapter 1).

Veatch found that production from an artesian well between tide levels on the coast at Huntingdon, Long Island, New York, fluctuated with the tides with a delay of about 2 minutes at low water and 8 minutes at high water (Veatch, 1906, pp. 10-13). The weight of the sea water clearly loaded the aquifer. This conclusion is not contrary to the conclusion reached in Chapter 1 (p. 17), that effective stress is independent of the depth of water over a sand. The significant parameters here are permeability and time: the tide rises and falls too quickly for equilibrium to be reached, and so some of the load is borne by the confined aquifer.

One of Veatch's more interesting observations (op. cit'. p. 24) was that the water-level in a 4.3 m (14 ft) well rose rapidly when it rained, but that this was not due to infiltration (it rose *before* the adjacent brook), nor to barometric or temperature changes. A similar, but reduced effect was observed in a 21.9 m (72 ft) well. He inferred that the rise was due to the weight of the rain-water on the ground. (If this seems unlikely, note that 1 cm of water on a hectare is 100 m^3, which weighs about

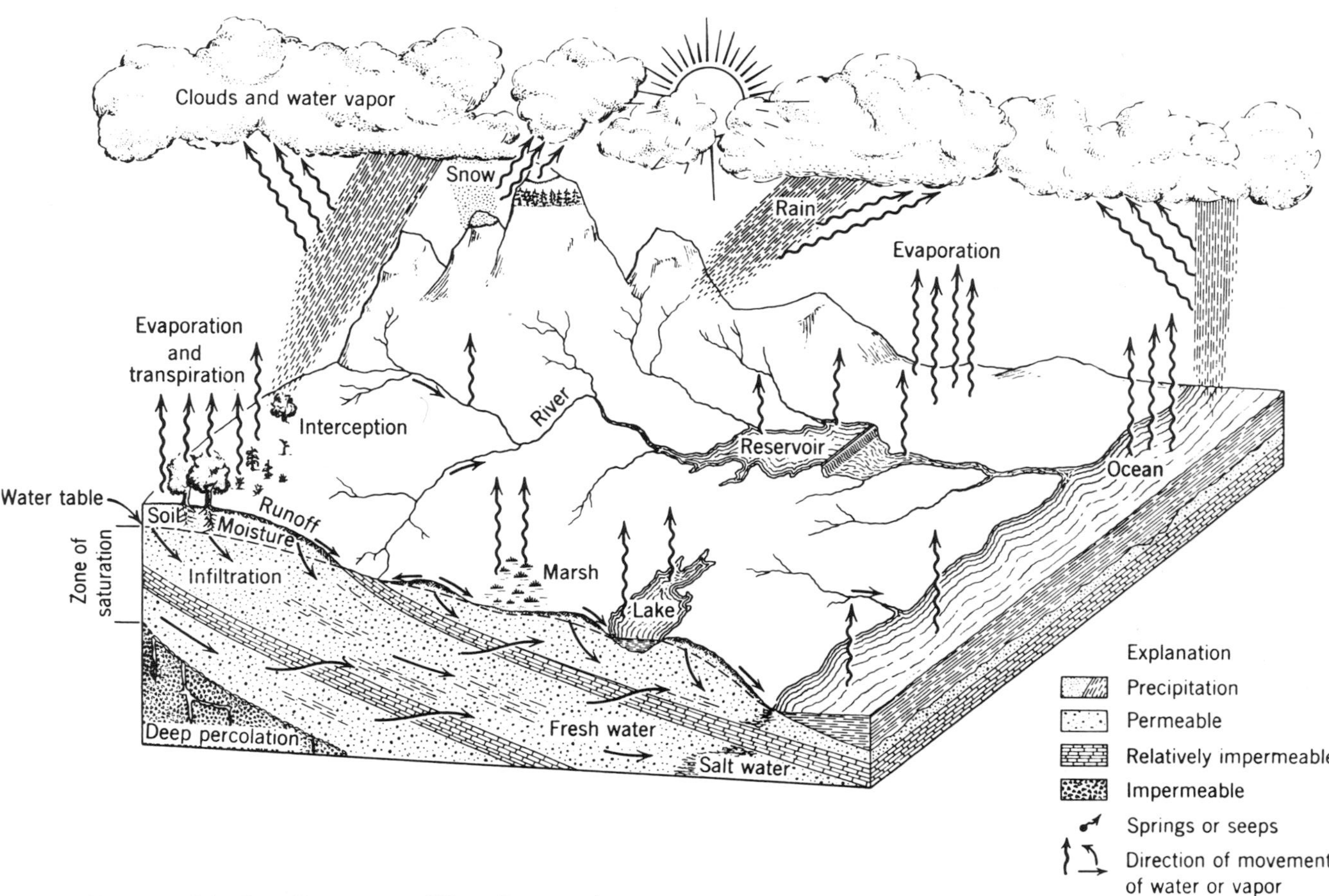

Fig. 5-2. (Courtesy of the Texas Department of Water Resources.)

100,000 kg – one inch on an acre weighs about 100 tons – and recall the effect of the freight train on the water-level in a well).

As an example of the early human influence on the relationship between an aquifer and a river, we may cite the well-known case of the river Thames and London's ground-water supply (Whitaker and Tresh, 1916; Buchan, 1938). London's main ground-water source has been the Chalk (Cretaceous) aquifer that underlies it. This is a synclinal artesian aquifer fed from intake areas over the outcrop north and south of the city, confined by relatively impermeable Tertiary beds above, and the Gault Clay (Cretaceous) below (Fig. 5-3). As London's demand for water grew with the city, so the pressure head in the wells declined (Buchan, 1938, pp. 28, 32) and the aquifer's potentiometric surface was lowered: the aquifer was produced for years beyond its *safe yield*. In the Thames estuary on the down-stream side of London, the Chalk crops out on the river bed over a distance of some kilometres. In earlier times, this area acted as a leak from the aquifer, with fresh water entering the river. Springs could be seen at low tide. With the decline in head, the flow here was reversed, and the aquifer was contaminated with saline water (at best!). This consequence was predicted by the Local Government Board reporting to Parliament in 1899 (See Whitaker and Tresh, 1916, pp. 53-54)*.

The normal interrelationship between aquifer and river flow was nicely studied by Ineson and Downing (1965) in two English rivers, the Itchen and the Stour. Figure 5-4 shows the Itchen hydrograph they obtained. Storm-water runoff accounts for the peaks: but the general pattern of flow corresponds to the seasonal water-table

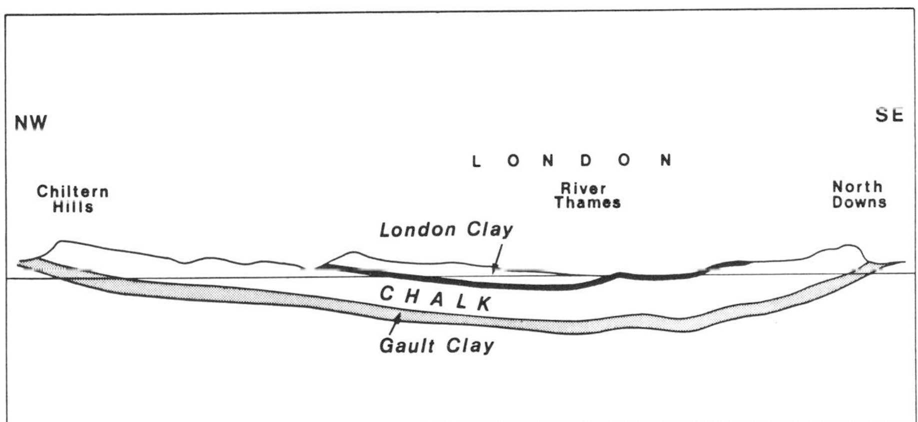

Fig. 5-3. Simplified geological cross-section through the Chalk aquifer beneath London, England.

* In the Arabian Gulf (I was told some years ago) the pearl fishermen were able to stay at sea for protracted periods because they collected fresh water from submarine springs. They would dive down with a jar, or skin, invert it over the spring, then close it. The growth of Bahrain and the concomitant increase in demand for water depleted these aquifers so much that the sea-water entered where fresh water had emerged.

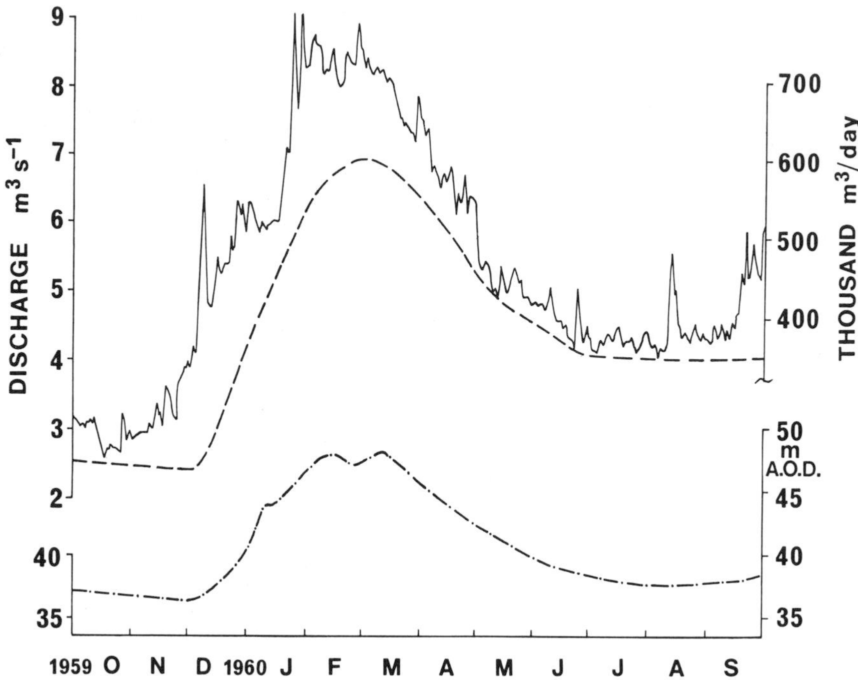

Fig. 5-4. Hydrograph of the river Itchen, England, and (below) water-levels in a near-by well (after Ineson and Downing, 1965, fig. 1).

fluctuations, as shown by the levels in a nearby water well. They estimated that between 75% and 85% of the flow in these rivers was due to ground-water flow; and we may take these figures as representative of rivers in temperature climates.

We shall now look at some further examples that illustrate the variations on the theme of water flow in nature, with special reference to the aquifer.

WATER FROM THE DANUBE TO THE RHINE

There can be few more impressive interactions between rivers and ground water than the transfer of water from the Danube to the Rhine through the Malm (Upper Jurassic) limestone aquifer (Fig. 5-5). The loss of Danube water into the '*brühl*' between Immendingen and Möhringen has presumably been known for some centuries: in 1719 Breuninger published an account of the head-waters of the Danube with the thesis that the water that entered the sink-holes in the Danube between Immendingen and Möhringen emerged as the spring at Aach, the largest in Germany, and so reached the Rhine.

The loss of water from the Danube was, of course, important to the towns and

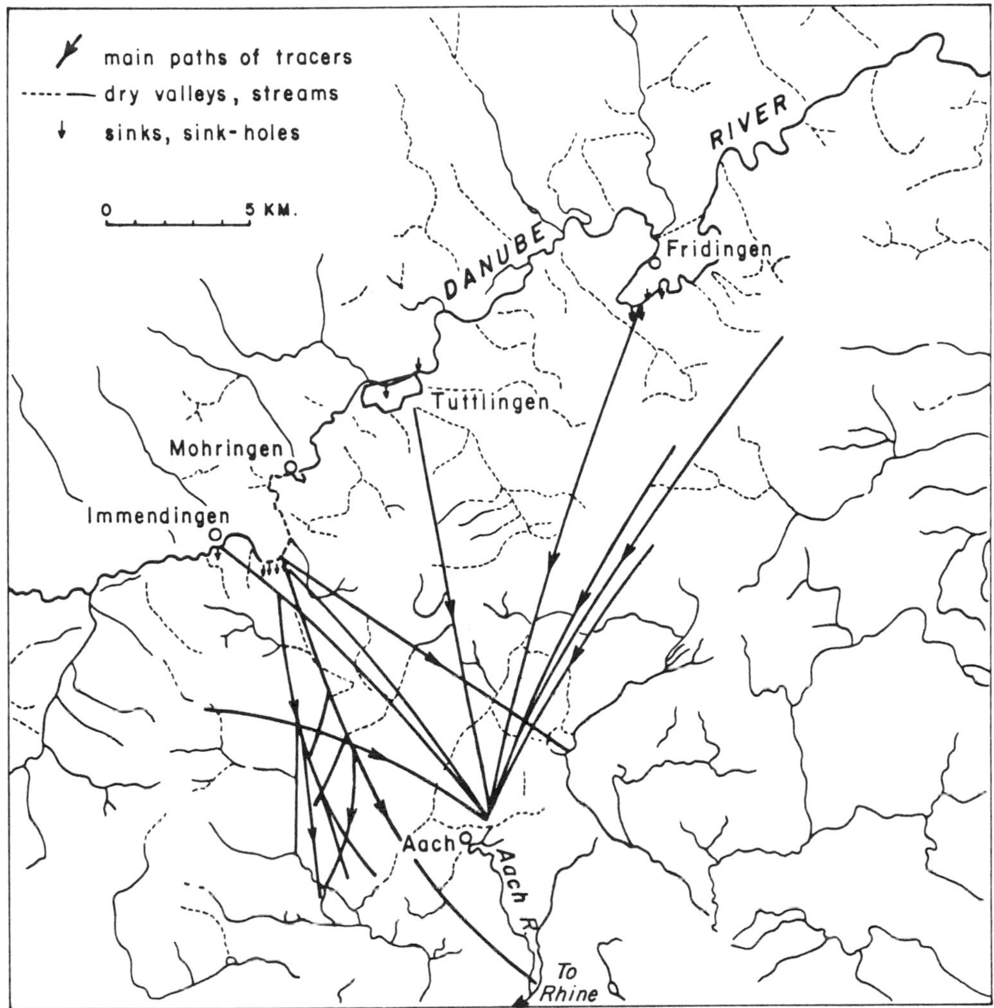

Fig. 5-5. Subsurface flow directions in area between the Danube and the spring at Aach (after Käss and Hötzl, 1973).

villages below the sink, particularly when at unusually low flow rates in the headwaters all the water disappeared underground (as happened in 1874 for the first time, it is said). This led in 1877 to what was perhaps the first use of tracers in ground-water studies*, when an oil tracer was introduced into the water at the sink and recovered at the spring above Aach (see Käss, 1969, 1972; Käss and Hötzl, 1973). The oil was introduced in the afternoon of Saturday, 22 September 1877, and

* An earlier inadvertent use of tracers occurred in the Teutoburger Wald, central Germany: two ducks vanished down a sink near Neuenbecken, to emerge after some days about 5 km away at the source of the Lippe!

from 6 am on the Monday morning, 24 September, the spring water at Aach had a distinct smell of creosote, which lasted about 6 hours (Knop, 1878, reported in Käss, 1969 – the year of the experiment is wrongly recorded here as 1878). At 4 pm on Tuesday, 9 October 1877, an organic dye sodium fluorescein (uranin) was introduced into the water at the sink, and recovered at the spring above Aach at dawn on 12 October.

The number of days per year during which the Danube ran dry between Immendingen and Möhringen, all the water being lost to the sinks, had been increasing during the period 1885 to 1929 at an average rate of 2.55 days/year (Wolf, 1931, reported in Käss, 1969). It seemed entirely reasonable to attribute this to the enlargement of the joints and fissures by solution of the limestone. However, Käss (1969, p. 243, Fig. 5-6) brought unpublished data of Pantle up-to-date, and the seven-year average of days of total loss per year shows a peak about 1950 of a little over 200 days/year, with a decline after that. This decline suggests a cause outside the aquifer, perhaps in the long-term weather changes. It also suggests that 1874 is unlikely to have been the first year in which the Danube ran dry between Immendingen and Möhringen.

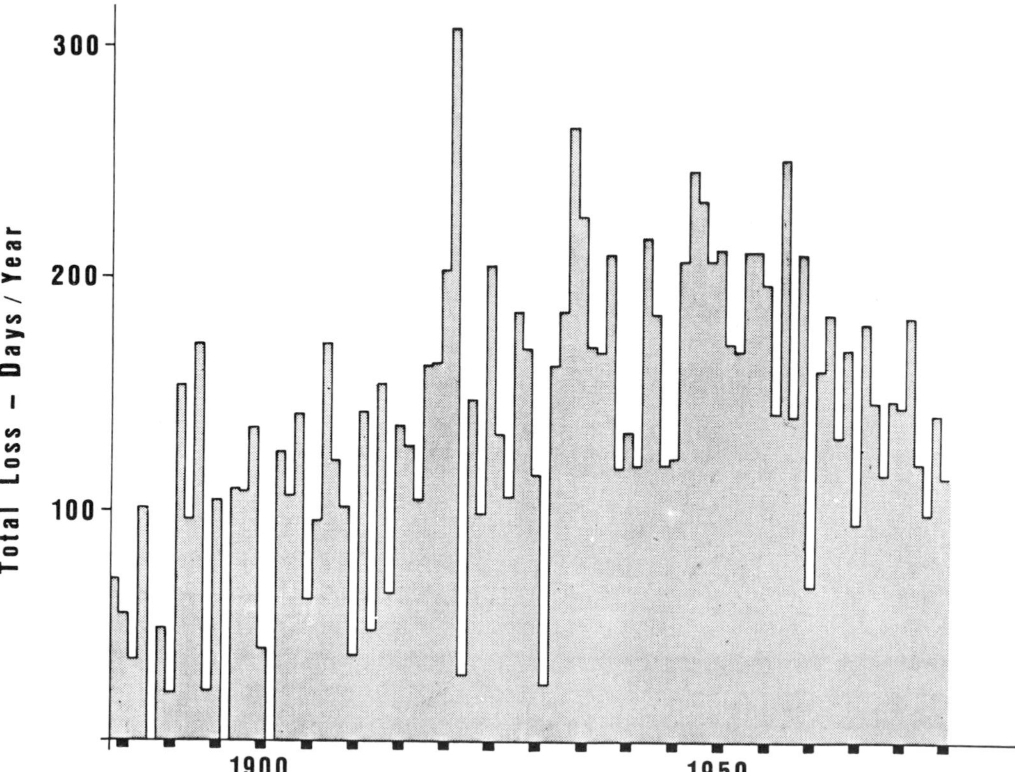

Fig. 5-6. Number of days the Danube ran dry at Immendingen each year since 1884 (figures by courtesy of Dr W. Käss).

In recent years, the flow of water to Aach has been amply demonstrated by a dozen tracers, not only from the Immendingen-Möhringen area, but also from further downstream at Tuttlingen (Duttlingen of Breuninger, 1719) and Fridingen. We have no means of knowing the precise paths of the water; but the spring at Aach acts as a major sink to the aquifer, with water flowing to it over an arc of about 90°. The volumetric rate of flow to the Aach spring, estimated to average 6 $m^3 s^{-1}$, varies according to the conditions – as does the velocity. A velocity of 0.035 $m s^{-1}$ was recorded in October 1971 (106 hours to traverse 13.3 km straight-line distance); and in August 1969 the recorded velocity was 0.070 $m s^{-1}$. The 1877 experiments indicate about 0.09 $m s^{-1}$ in September, and about 0.06 $m s^{-1}$ in October. Immendingen is close to 650 m above sea-level: the spring at Aach, 467 m above sea-level. The hydraulic gradient is therefore about 0.014. It is hard to understand how these ranges of velocities occur when the constraints seem to be so limited (Darcy's law does not apply to flow through fissures, but the laws for pipe and channel flow are, as we have seen, very similar with similar constraints).

The Malm aquifer extends over a wide area of the eastern Molasse basin (at least) and is of some interest. Along the southern outcrop of the Malm (Fig. 5-7) the water-table elevation is given approximately by the level of the Danube: at the south-western end, the water-table is not higher than 650 m at Immendingen or lower than 467 m at the Aach spring. To the south-east of Aach, the water-table elevation appears to be limited by the Bodensee, 395 m above sea-level, although the Malm does not crop out around it. Drilling for petroleum has provided other data points by virtue of pressure and depth measurements. Figure 5-7 shows the potentiometric-surface map of the Malm aquifer in the Molasse basin drawn by Lemcke (1976, p. 11, fig. 1). The most striking feature of this is the apparent sink in the aquifer east of Munich (München). If the aquifer is in physical continuity over this area, we infer that water is not only flowing from the Danube to the Rhine over a wider area than from near Immendingen to Aach, but also flowing into the deeper part of the basin.

The point of interest here is that the elevations of the potentiometric surface indicated by the borehole data are considerably less than the ground-surface elevations (around 600 to 700 metres above sea-level) and there is no obvious surface outflow from this sink. Other aquifers in the Molasse basin also have pore-water pressures well below the normal hydrostatic pressure.

If this were the only case of its sort, it might seem that there must be an error in the data or its interpretation. Figure 5-8 shows the potentiometric-surface map of a Mesaverde sandstone in central U.S.A., and Figure 5-9 that of the Viking Sandstone in Canada (Hill et al., 1961). Perhaps they are right in suggesting osmosis as the cause.

If liquids of different salinities are separated by a semipermeable membrane, the less saline solution moves through the membrane into the more saline solution; and if equalization of the salinity cannot be attained, a pressure difference results. This process is called *osmosis*. If the sub-normal pressures are one side of this coin, the other has yet to be found.

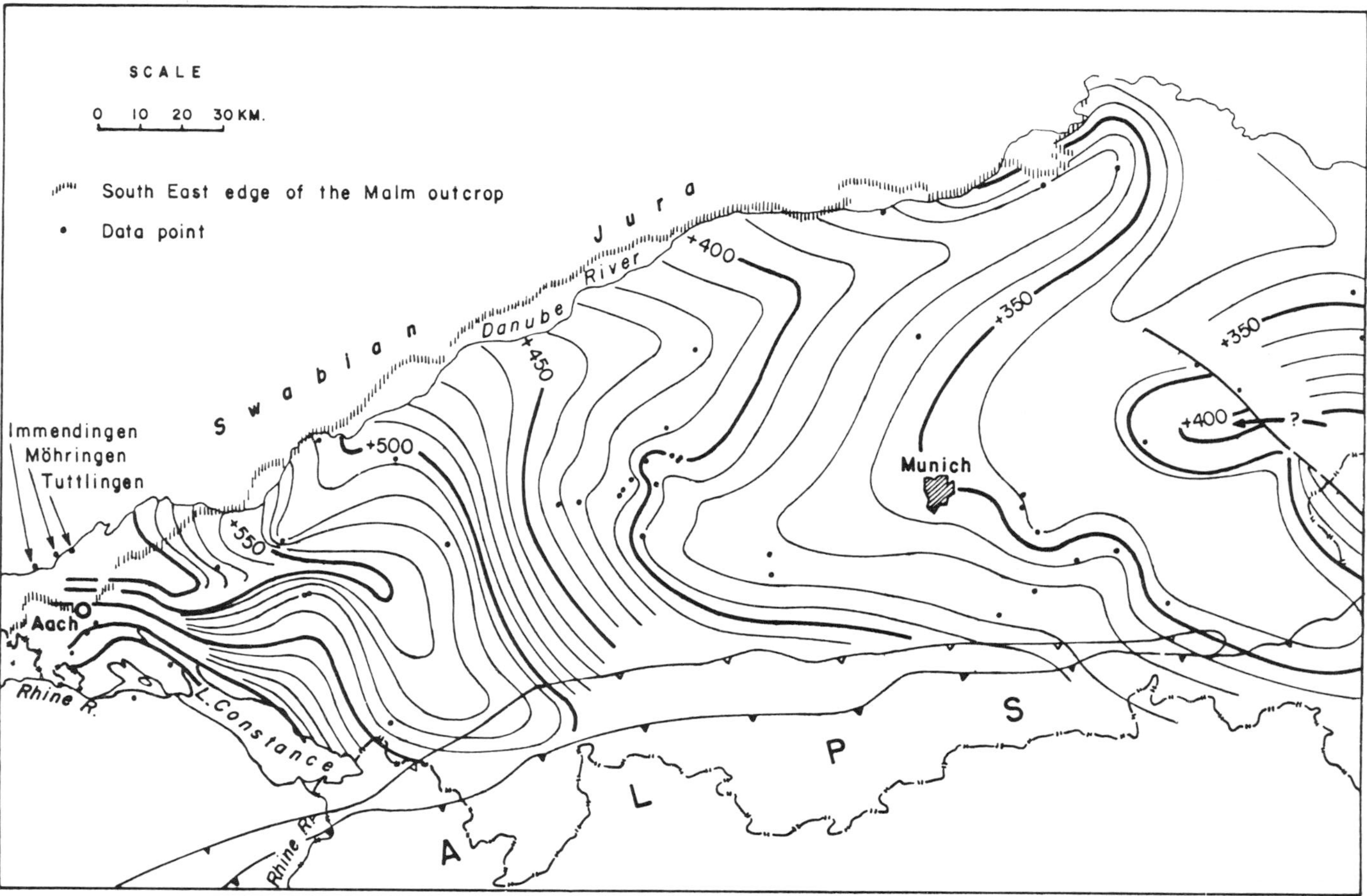

Fig. 5-7. Map of potentiometric surface of the Malm aquifer in the Molasse basin of southern Germany (after Lemcke, 1976, fig. 1).

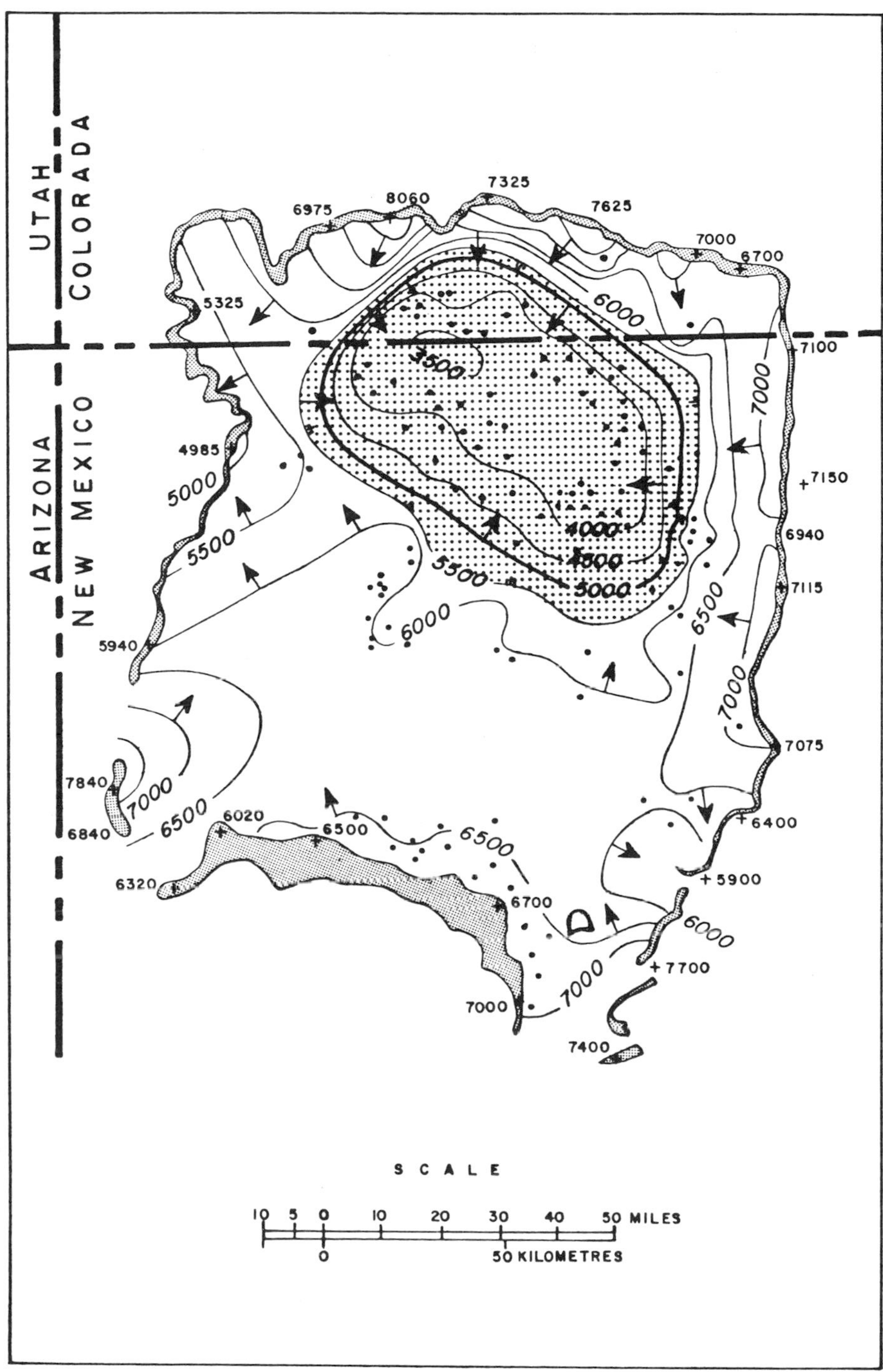

Fig. 5-8. Map of potentiometric surface of a Mesaverde sandstone, central U.S.A. (after Hill et al., 1961, fig. 3).

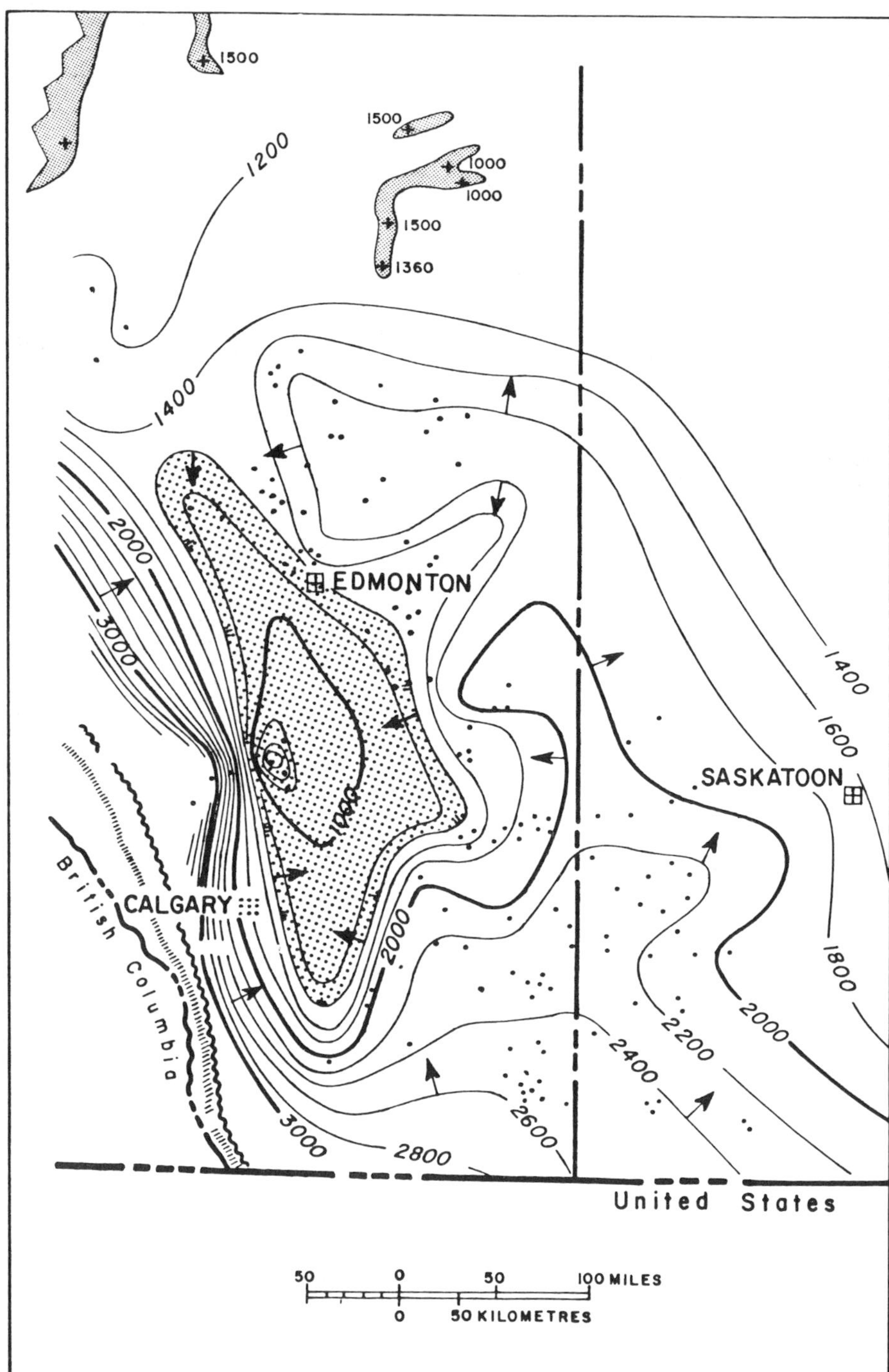

Fig. 5-9. Map of potentiometric surface of the Viking Sandstone in Canada (after Hill et al., 1961, fig. 2).

OCEANIC ISLANDS AND COASTAL AQUIFERS

The behaviour of coastal aquifers is important to many countries because of population concentrations and their demands for water for industry, agriculture, and domestic use. Perhaps no country is so conscious of the sea as The Netherlands, nor as conscious of the proper use of fresh water. Levels are important to the Dutch because 'downhill' does not lead very far. It is hardly surprising, therefore, that it was in Holland that it was first noticed that water wells encountered the water-table not at sea-level but some distance above it. In a long note, Drabbe and Badon Ghijben (1889)* published details of water-levels near Amsterdam, noted that the water-table was commonly above sea-level, and proposed the explanation that the denser sea water supports a longer column of the less dense fresh water. They took the specific gravity of Zuiderzee water (then open to the sea) to be 1.0238 and concluded that if a is the elevation of the water-table above sea-level, the depth of 'balance' between the two waters is given by $a/0.0238 = 42a$ (Drabbe and Badon Ghijben, 1889, p. 21).

Some years later, Herzberg (1901), working on the North Sea coast of Germany, found the same features and drew the same conclusions.

Briefly stated, their conclusions were that the ratio of the elevation of the water-table above sea-level to the depth of the fresh/salt water interface below sea-level is equal to the ratio of the difference between the densities of the sea water and fresh water to the density of the fresh water, i.e.,

$$h/z = (\rho_{\mathrm{sw}} - \rho_{\mathrm{fw}})/\rho_{\mathrm{fw}} \simeq 1/40. \tag{5.1}$$

This has become known as the Ghijben-Herzberg relationship or 'law' (one wonders why Drabbe was left out, and Herzberg put in).

The Ghijben-Herzberg relationship implies static equilibrium between the lens of fresh water and the sea water: that the mass or weight of fresh water in a column $h + z$ is equal to the mass or weight of sea water displaced, or

$$(\rho_{\mathrm{fw}} - \rho_{\mathrm{air}})gh + (\rho_{\mathrm{fw}} - \rho_{\mathrm{sw}})gz = 0. \tag{5.2}$$

Drabbe and Badon Ghijben probably understood that this is unstable, but Versluys (1919) pointed it out clearly. He went further and argued that the consequences of

* This reference is almost invariably given as Badon Ghijben, 1889, or Ghijben, 1889. W. Badon Ghijben was in fact the junior author (Badon is part of the surname). There are several matters of interest in this note: for example, they clearly understood the significance of the sloping water-table because on p. 16 they wrote 'Neemt men in aanmerking, dat de grondwaterstand in de duinen in het algemeen de golvende zandoppervlakte op eene geringe diepte volgt en *hier* dus zeker *verscheidene* meters boven den waterspiegel der Zuiderzee is verheven, dan schijnt het zeker, dat er in den ondergrond een voortdurende stroom moet zijn *van* de duinen *naar* de Zuiderzee' – 'When one considers that the ground-water level in the dunes follows in general the wavy surface of the sand at shallow depth and thus *here* is certainly elevated *several* metres above the level of the Zuiderzee, then it appears certain that there must be a continuous underground flow *from* the dunes *to* the Zuiderzee' (their italics).

this instability were that the relationship was much more complicated, and that the fresh water flowed into the sea over a zone, as in Figure 5-10. He clearly understood that the fresh water is not at uniform potential* in the lens, so it is in a state of constant motion; and if no rain rell, the lens would dissipate by flow to the sea near sea-level.

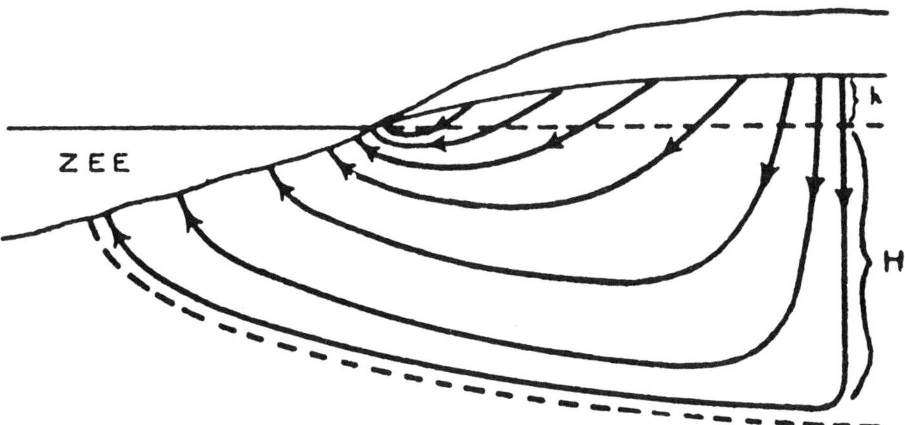

Fig. 5-10. Versluys' diagram showing flow to the sea in a fresh-water lens under coastal dunes (Versluys, 1919, fig. 4).

The Ghijben-Herzberg lens is usually shown diagrammatically as in Figure 5-11, but there is great distortion in such diagrams due to the grossly exaggerated vertical scale. This falsely suggests that the water flow is mainly vertical. Figure 5-12 shows a half-lens on a more natural scale.

The shape of the water table indicates that the water in the lens is flowing away from the centre line, in general (as Drabbe and Badon Ghijben appreciated), and considerations of symmetry require the flow at the centre line to be vertically downwards initially (as Versluys appreciated). The matter of equipotential surfaces in 2-fluid systems was treated by Hubbert (1940, pp. 868-870, 924-926), the essence of which is this: if the sea water is static and the fresh water flowing, the sea water is at constant potential so the fresh-water/salt-water interface is an equipotential surface of the sea water. On the fresh water side of the interface, the equipotential surfaces of the fresh water are normal to the interface because it is a boundary to the flow. If Δh is the vertical interval of equipotential contours on the water-table, $\Delta h \rho_{fw}/(\rho_{sw} - \rho_{fw})$ is the corresponding vertical interval of equipotential contours on the interface. The equipotential surfaces between the corresponding contours are curved, meeting the interface at right angles; so the interface is displaced outwards from the centre line from the position predicted by the Ghijben-Herzberg 'law', and is therefore deeper than that prediction.

* Versluys (1917, p. 24, eq. 6) had already explicitly defined 'hydrostatic potential', or 'potential' for short, as the sum of the elevation and pressure heads. This differs from Hubbert's fluid potential only in the factor g (Hubbert, 1940, p. 802).

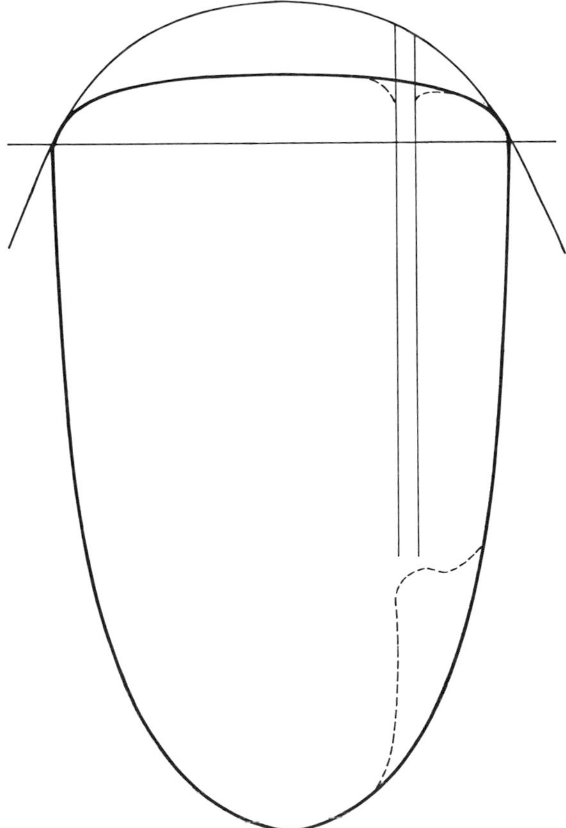

Fig. 5-11. Schematic section through Ghijben-Herzberg lens in an island, vertical scale grossly exaggerated. Sketched in is the influence of a producing water-well.

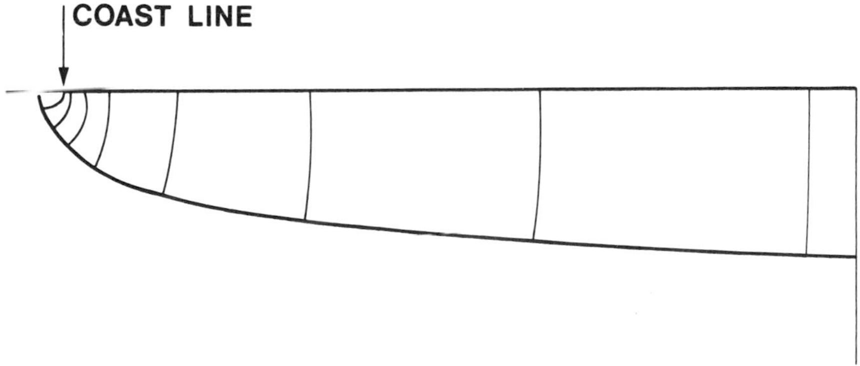

Fig. 5-12. Half-lens at more natural scales, with equipotential lines.

If the slope of the water-table between two equipotential contours is a below the horizontal in the direction of flow, and the material is isotropically permeable, we have from Darcy's law

$$\sin a = \frac{\Delta h}{\Delta l} = \frac{q}{K},$$

where Δl is the corresponding length along the watertable. At the interface, the flow is inclined upwards: the flow is converging. The same quantity of water crosses each equipotential surface in the fresh water, so q ($= Q/A$) increases towards the coast, and the slope of the watertable increases towards the coast. By a similar argument, the slope of the interface above the horizontal also increases towards the coast. The surfaces bounding the fresh water are curved, concave to sea-level (as Versluys appreciated).

In practice, it is so undesirable to drill water wells to produce from near the edge of the lens that its precise shape is not very important, and the Ghijben-Herzberg 'law' can be taken as a limit. In any case, there are many disturbing influences – irregular rainfall, tides, anisotropic porous solids, and perhaps diffusion at the interface with the salt water*. The effect of producing water from a well in a Ghijben-Herzberg lens is to superimpose another field of flow. The field imposed by a producing well far from the coast is much stronger than the natural field, and it will be appreciated that a small apparent depletion that lowers the water-table slightly has a far more serious hidden effect at the bottom of the lens (sketched on Fig. 5-11).

The Pacific islands are particularly dependent on their Ghijben-Herzberg lenses because they, like all oceanic islands, are entirely dependent on rainfall for water. The lens is their natural storage reservoir for times of drought. The Second World War imposed enormous strains on many of these resources.

The island of Oahu in Hawaii, on which Honolulu and Pearl Harbour are situated, is an interesting example of a Ghijben-Herzberg lens because erosion has led to the accumulation of a relatively impermeable layer of sediment over the zone of natural seepage at and below sea-level (Fig. 5-13). This caprock, by distorting the Ghijben-Herzberg lens, gives to it a local artesian quality. According to Ohrt (1947) and to Wentworth (1952) the (total) head above mean sea-level in a well drilled through the caprock in the central Honolulu area was about 12.8 m (42 ft) in 1880, implying by the Ghijben-Herzberg relationship a depth below sea-level to the interface of 486 m (1600 ft). Wells drilled here to 366 m (1200 ft) below sea-level produced fresh water. The uncontrolled drilling of artesian wells had led to a serious decline in the total head by 1910; and by 1926 it was down to 7.6 m (25 ft) when a conservation programme, aided by rainfall, slightly reversed the decline. By 1947,

* Versluys (1919) argued persuasively that the brackish zone near the interface is the result not so much of diffusion but rather of incomplete washing by fresh water, because the salinity does not increase with depth throughout the lens, but only in a transition zone near the interface.

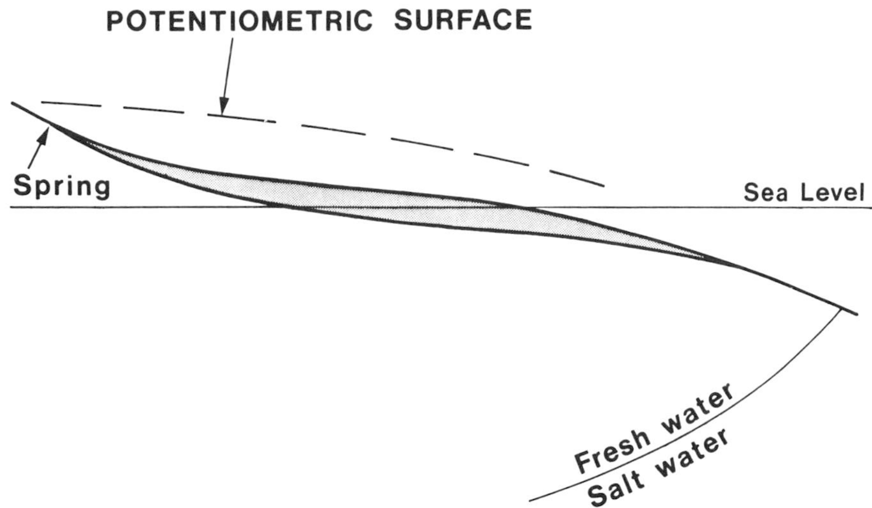

Fig. 5-13. Schematic section showing artesian zone on a low-lying coastal area covered by relatively impermeable sediments.

nearly 500 artesian wells had been drilled through the caprock and, in spite of the strains of war, the total head was only slightly below the 1926 level.

The loss of fresh water in storage in the Ghijben-Herzberg lens was therefore enormous, estimated at 40% of the natural quantity, and it is generally considered that such losses are irrecoverable in the time-scale of our lives. The expansion of the lens probably leaves a thicker transition zone of brackish water. In the central Honolulu area, salt water has encroached to 305 m (1000 ft), with local indications of encroachment to 245 m (800 ft).

Ghijben-Herzberg lenses are not the only source of fresh water in volcanic oceanic islands. Perched aquifers may be contained between dykes, and water may flow down old lava channels, contained by relatively impermeable tuffs. On the Atlantic island of Tenerife (Canary Islands, Spain) the main water production comes from *galerías*, tunnels about $1\frac{1}{2} \times 1$ m dug into the hillside with a slight upward slope. These penetrate progressively older strata, and so may encounter percolating water or a perched aquifer. These *galerías* are purely speculative ventures because water is not known to exist along their paths beforehand. Some are about 4 km long, and represent years of work and investment. Naturally, the failures tend to be longer than the productive ones. Around 1960, 98% of Tenerife's water supplies came from *galerías:* 162 million cubic metres a year from 386 producing *galerías* with a total length of 763 km (*Canarias. anexo al Plan de Desarrollo Económico y Social* 1964-1967: Presidencia del Gobierno, Edición del Boletin Oficial del Estado, 1964, pp. 65, 68, 72). These figures suggest an average length of about 2 km.

ARID REGIONS: THE QANAT OF IRAN

In arid regions, as Sandford (1935) pointed out, the rainfall may be slight but it is not everywhere insignificant. Working in the desert to the west of the Nile, around the common borders of Libya, Egypt, Sudan and Chad, he measured the elevation of water wells with an aneroid barometer and the depth to the water table, and so produced a contour map of the water table (Fig. 5-14). The shape of the water table indicates a southerly source for the ground water, with flow to the west south of Tibesti, and to the north-east to the Nile. It is the development of aquifers in arid regions that is of interest rather than the aquifer itself, and the most important method of development is due to the Persians.

Throughout the area of the former Persian and Arab Empires, ground water is developed for agricultural and domestic use by a system of wells joined together by a tunnel or gallery that brings water from the foothills to the plains (Fig. 5-15). The system is of great antiquity, for it had been developed by the year 800 BC, and was introduced into Egypt in 512 BC (while Egypt was a Persian province, during the reign of Darius) in the oasis of Kharga – to such effect that the event was recorded by an inscription (Butler, 1933).

Essentially the same system is found from Morocco to Afghanistan, and in southern Arabia around the Oman mountains and in the Yemen. Different names are used in different parts: *shat-at-ir* in Morocco, *foggariur* in Algeria, *qanát* or *kanát* in Iran and Baluchistan, *kariz* in Iraq and Afghanistan, *falaj* (plural, *aflaj*) around the Oman mountains, and *shariz* in the Yemen. They are of interest because they illustrate a practical understanding of ground-water hydrology amongst the ancients that is still of inestimable value. We shall call them *qanat*, generally, as they are called in their country of origin. They do in arid climates by the labour and ingenuity of Man what rivers do in more humid climates.

The first stage in the construction of a qanat (Noel, 1944; Beckett, 1953) is the sinking of exploratory and appraisal wells in the area thought to contain a suitable aquifer that could supply water to a settled area (or land that could be settled). These are hand-dug, about 0.75 m in diameter, and may be about 100 m deep. When satisfied that there is sufficient water, a survey is made to determine the level at which the qanat and the water will emerge. The traditional method of survey is to lower a line down the 'mother well' (the one furthest away from the area of intended use) and to tie knots at the ground surface and the water table. A line is then strung from the ground surface at the well to the top of a pole sited in the direction in which the qanat will run. This line is levelled by wetting it and adjusting the level at the pole until the water drops collect at the middle of the line. The difference in elevation is then subtracted from the cord that records the depth to the water table, and marked with another knot or pin. This process is continued until the loss of elevation equals the original length between the knots (illiteracy is not to be confused with ignorance!). The gradient of the qanat is then decided (usually about 1:1500).

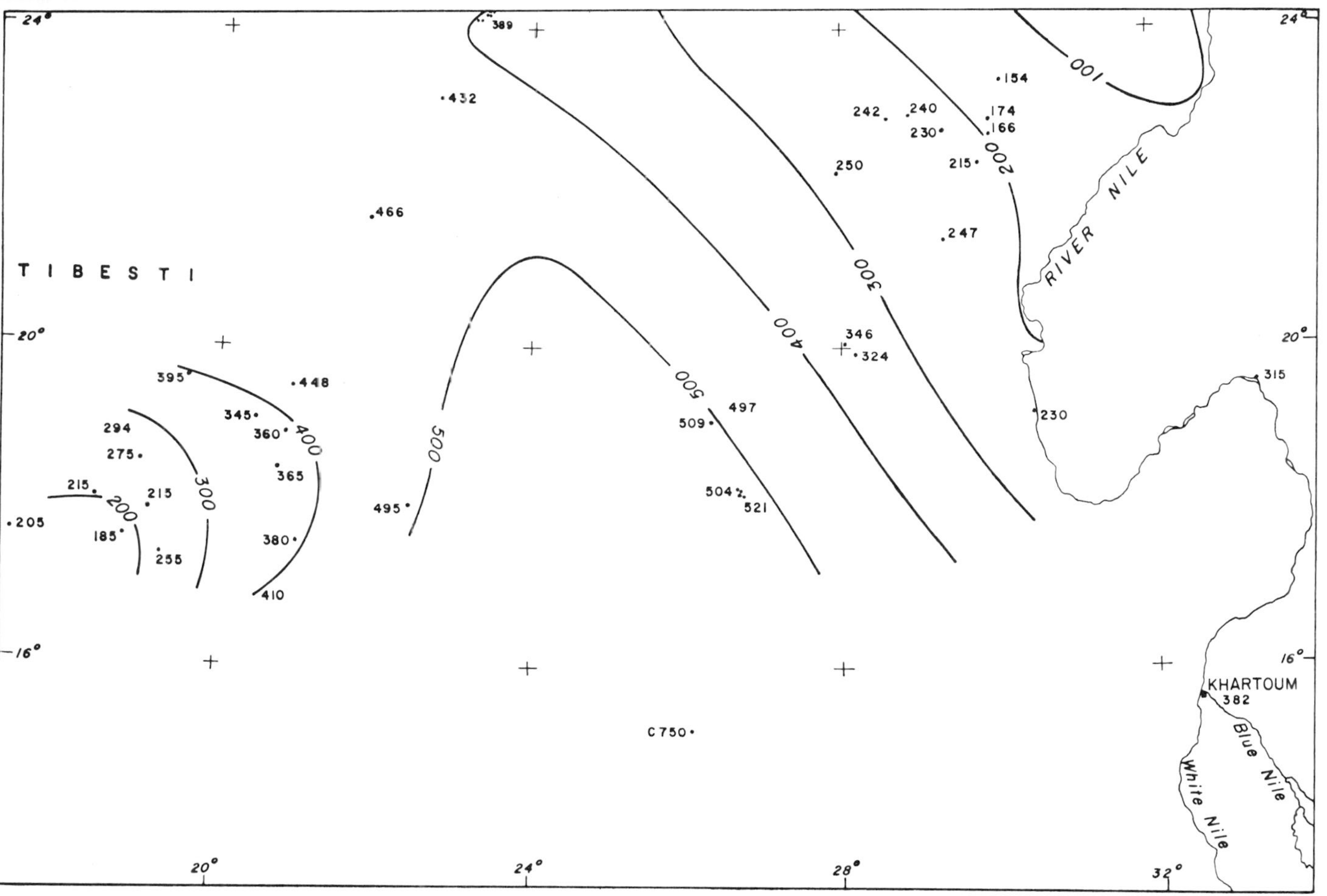

Fig. 5-14. Map of water-table in Nubian Sandstone in desert west of the Nile, elevations in metres (after Sandford, 1935).

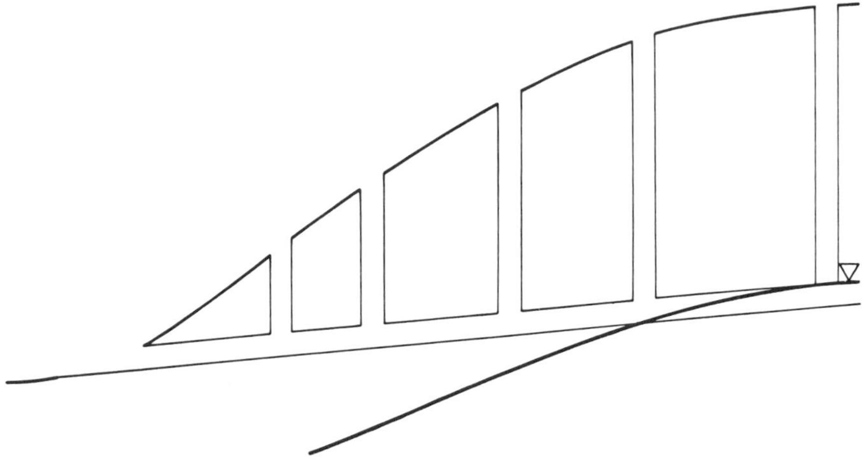

Fig. 5-15. Schematic section through a qanat.

Shafts are dug to the required level along the planned route of the qanat, every 300 m or so, and the construction of the qanat itself is begun in the dry section, working from the lower end. Shafts are dug at intervals of about 25 m where the depth is slight, for ventilation and the extraction of spoil. The interval is increased as the depth becomes greater. The qanat itself is dug from shaft to shaft: it is about 1.25 m high and 0.75 m wide, and Noel (1944) reports that these dimensions are considered to be the largest consistent with safety. Two lamps are used to keep the qanat on course (as they are in the galerías of Tenerife). Soft sections are lined with oval rings of baked mud, straw, and dung. Once the dry section has been completed, construction into the aquifer can be started.

Production usually declines rapidly at first, but if the aquifer is good, soon stabilizes. Qanat may be deepened to offset decline in production. Beckett (1953) and Smith (1953, p. 133) reported that a qanat near Kirman in SE Iran had its yield stimulated by an earthquake, the source being changed from a seepage to a spring. Ideally the water flows at $\frac{1}{4}$ to $\frac{1}{2}$ m s^{-1}. At greater rates of flow, erosion of the qanat may occur, leading ultimately to its collapse.

Small fish swim in all but the saltiest qanat, and are said to have come from the rocks. They are also found in the aflaj of Buraimi in southern Arabia, and the same explanation is given there. Smith (1953, p. 81) ascribes their presence to their being washed in from rivers in flood; but fish have been reported from water wells in many countries. Lyell (1867, p. 393) reports live fish from artesian water wells from 47.5 m in Germany and 53.3 m in the Sahara: Norton (1897, pp. 167-168) reports live crustaceans from an artesian well in Texas, and live fish from wells near Aberdeen, Dakota, U.S.A. Smith (1953, p. 135) collected one insect larva from the source of a qanat.

Butler (1933) records that the water supply for the entire city of Tehran came from

36 qanat, ranging in length from 13 to 26 km and passing in places more than 150 m below the surface. These supplied 0.17 $m^3 s^{-1}$ in autumn. A qanat near the foot of *Kuh-i-Kurgis* was said to be 90 km long, while lengths of 40 to 45 km are not uncommon. The yield from a qanat may be uniform throughout the year, or show seasonal variations. Thesiger (1959, p. 290) found during his 1948-49 journey through southern Arabia that some villages in the southern Oman mountains at the northern end of the Wahiba sands had been abandoned because their aflaj (as the qanat are here called) had run dry. His description indicates that these aflaj did not reach the surface: they were probably like those in Iran illustrated by Graves (1975, p. 43). Further to the north-west from the villages visited by Thesiger, on the west flank of the Oman mountains, the villages of the Buraimi oasis derive most of their water from aflaj that pass into the foothills of the Oman mountains. These were collectively producing about 0.4 $m^3 s^{-1}$ in the mid-1960s.

The question arises, how does the water reach the land surface in a qanat? The short answer is that the loss of head incurred in channel flow is very much less than that in the aquifer, so that if the aquifer somewhere has its water table above the level of the land to be irrigated, it can be exploited. That part of the completed qanat that is in the aquifer is a sink to that aquifer, and water flows into the qanat down the potential gradient that is set up by its construction (Fig. 5-16). The large surface area exposed to the aquifer accounts for its productivity, even when the permeability is low. The reader will recognise that in a transverse section through the qanat in its producing region, there will be a drawdown analogous to that of a producing water well; but the qanat is a sub-horizontal line sink. The relatively high initial produc-

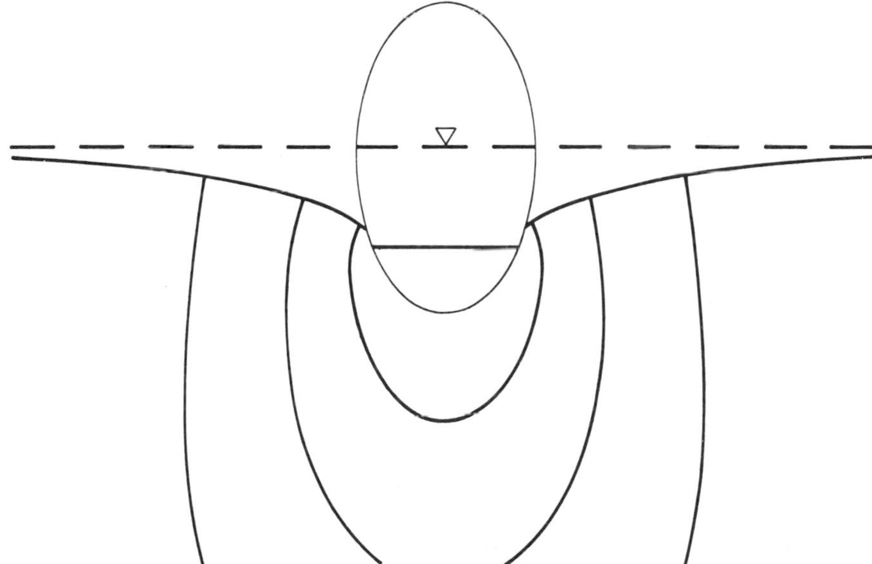

Fig. 5-16. Schematic cross-section through a qanat where it penetrates the aquifer, showing equipotential lines from which radial flow is inferred.

tion, and rapid decline, is due to the production of the volume of water during drawdown before steady flow has been achieved.

Clearly the flow is not like a fully-penetrating well, but like the flow to a perennial river (discussed on p. 90). The potential in the qanat at any point is the lowest in the cross-section through that point, so there is radial flow within the aquifer to the qanat, normal to the equipotential surfaces that pass under the qanat. The rate of flow in the qanat increases downstream until the qanat leaves the aquifer.

If the construction of qanats could be automated, they could still serve the arid regions of the Earth.

THE GREAT ARTESIAN BASIN

The Great Artesian Basin (Figs. 5-17, 5-18, 5-19) lies under nearly two million square kilometres of eastern Australia, under parts of the States of Queensland, New South Wales, South Australia and the Northern Territory. It has many aquifer in five main groups, ranging in age from Triassic to Cretaceous, the sediments older than Cretaceous being dominantly non-marine. It owes its artesian quality to Tertiary and Cretaceous marine shale aquicludes, intake areas in the east elevated about 500 m above sea-level, substantial rainfall on the intake areas, and a topographic slope to the west of the intake areas that is steeper than the hydraulic gradient. Its importance for the rural development of interior eastern Australia is inestimable.

The first artesian well was drilled in 1887 near Cunnamulla in Queensland, after a geological report indicated the possibility of artesian supplies: it found a flow of 4.2×10^{-3} m^3s^{-1} at 393 m (Queensland, Artesian Water Investigation Committee, 1955). By 1914, 1229 wells had been drilled, producing a total of 18.5 m^3s^{-1}. Thereafter the total flow began to decline: by 1950, it was 12 m^3s^{-1}. The decline of head amounted to over 100 m in places, but the residual head averaged 15 to 30 m above ground surface. Between 1890 and 1950, the total volume of water withdrawn was about 3×10^{10} m^3 (30 km^3) from the Queensland portion, and the investigating committee (op. cit.) estimated that the final 'steady state' flow would be reached with a production of 11 m^3s^{-1} from 520 permanent artesian bores. Other wells would become subartesian and would have to be pumped.

The average depth to the aquifers in Queensland (if we take the average depth of bore holes drilled to 1953) was 439 m, with the shallowest being only 2 m and the deepest, 2136 m deep. Over 1000 km of artesian hole have been drilled; registered boreholes, artesian and subartesian, total over 2000 km.

As we saw in previous chapters, the total head of an artesian aquifer that is perfectly sealed is horizontal, at the elevation of the intake area: but any leakage leads to flow towards the leak, and a corresponding slope to the potentiometric surface. The Great Artesian Basin had many leaks before humans started making them. There were springs in the intake area (Whitehouse, 1955) as a result of erosion reaching the water table, as a result of bedrock reaching the surface, and less-readily definable springs known as *mound springs*. The mounds of mound springs are built from debris and sediment carried to the surface by the water, with a result similar to mud volcanoes. These mound springs also occur outside the intake area. They are of particular interest, and we shall return to them shortly. The natural leakage from the Great Artesian Basin was considerable, and so its potentiometric surface was not horizontal. Its form has been deduced from a simulation model by Habermehl (1980) and is reproduced here as Figure 5-17.

At Cunnamulla (elevated about 189 m above sea-level) the original natural loss of total head was about 137 m in 400 km (a gradient of 3.4×10^{-4}) towards the south-west from the intake area, leaving a total head of about 320 m above sea-level, locally about 130 m above ground level. We do know that water production has considerably altered the shape of the potentiometric surface. Near Cunnamulla, where there was early development of this ground water, the 1,050-foot (320-m) equipotential line retreated about 165 km northwards between 1899 and 1950, with a reduction of head of 122 m (Ogilvie, 1955, p. 46 and fig. 94). And Spring Creek (90 km south-east of Boulia) originally flowed for more than 130 km to the Diamantina River, the water emerging from springs at a temperature of 18°C to 38°C with 0.61 kg m^{-3} (about 600 ppm) dissolved salts. By 1896 it had been reduced to about 30 km of flow, and by 1950 it had ceased flowing.

The general form of the potentiometric surface in 1970 is shown in Figure 5-17 (general, because different aquifers have different heads or potentials), and this indicates that the regional flow patterns have not been changed, the main flow being westwards, then south-westwards to Lake Eyre (which is 12 m below sea-level) and beyond. In the north (off the map) flow is to the north under the Gulf of Carpentaria.

That there has long been natural leakage in the south-western part of the basin is shown by the presence of numerous mound springs. These mound springs build up a cone of sediment, debris, and salt *until they reach the potentiometric surface*, at which level all kinetic energy is, of course, lost. Whitehouse (1955) observed that there are extinct mounds in the far south-west, the tops of which are about 30 m above the level of the active mounds (one must assume that he was of the opinion that they could not build up a further 30 m).

The temperature of the artesian water varies over the basin, and the geothermal gradient varies from about 30°C/km to 80°C/km, the highest being apparently 166°C/km near Julia Creek. West of Winton, Queensland, 2.6×10^{-2} m^3s^{-1} of boiling water was produced from 1372 m, indicating a geothermal gradient of 55-58°C/km if the mean surface temperature is taken as 25-20°C. However, accurate temperature measurements of the aquifers are lacking. The gradient increases from east to west in Queensland.

The quality of the water is also variable (Fig. 5-18). Airey and others (1979), in a

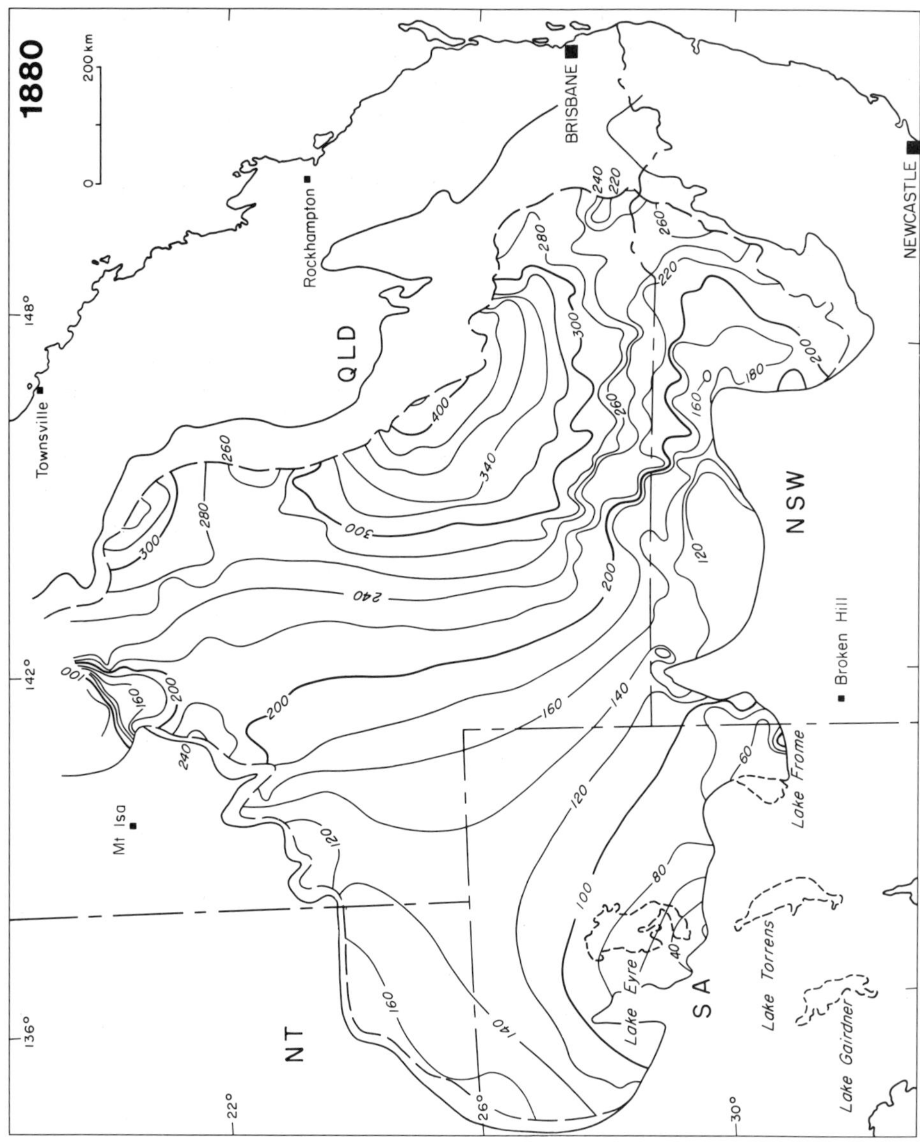

Fig. 5-17. Potentiometric maps of the main Lower Cretaceous-Jurassic aquifers for 1880 and 1970, modified from computer-simulated model results. Datum is mean sea-level, and potentials relate to pure water at 15 °C. (Habermehl, 1980, fig. 15, reproduced by courtesy of the Bureau of Mineral Resources, Canberra.)

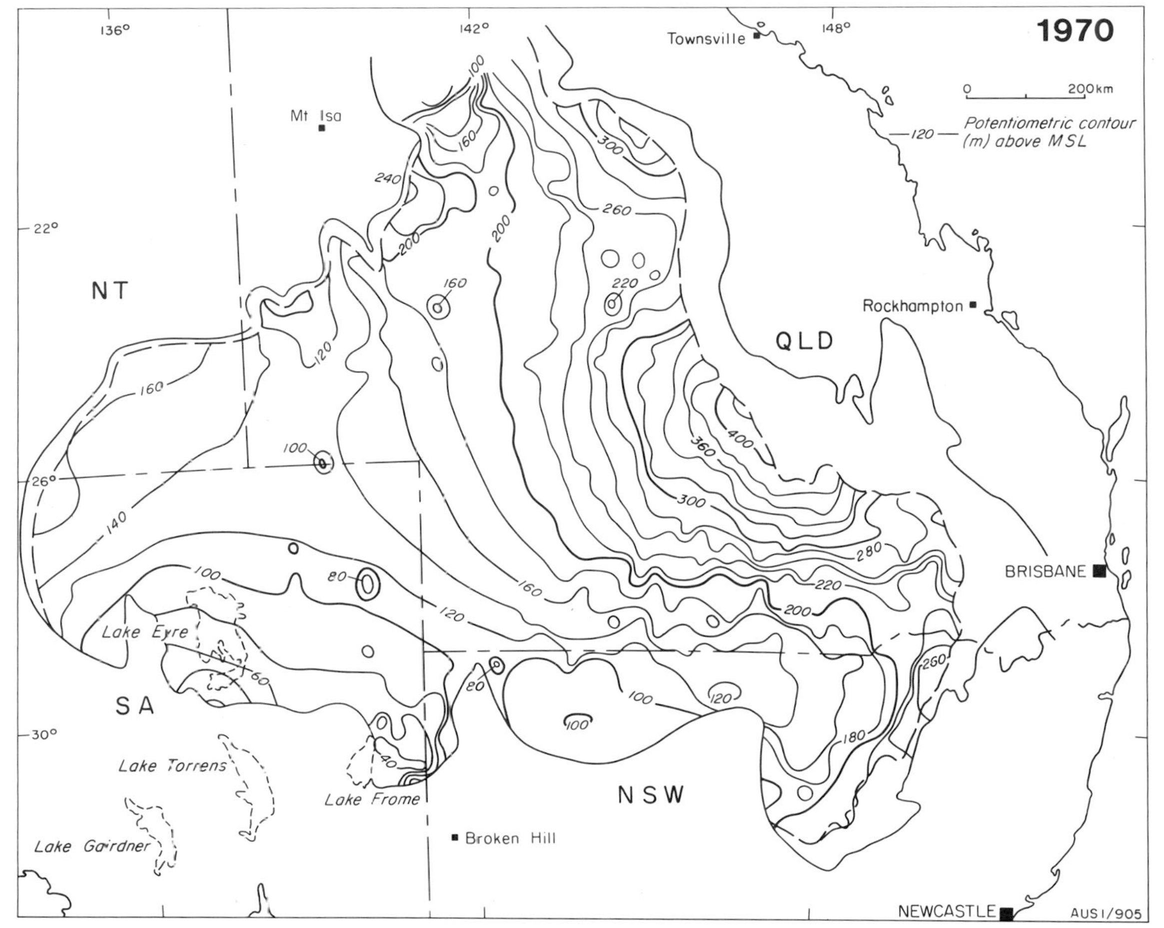
1970
0 200km
—120— Potentiometric contour (m) above MSL
Townsville
Mt Isa
Rockhampton
QLD
NT
BRISBANE
SA
NSW
Lake Eyre
Lake Torrens
Lake Gairdner
Lake Frome
Broken Hill
NEWCASTLE
AUS1/905
136°
142°
148°
22°
26°
30°

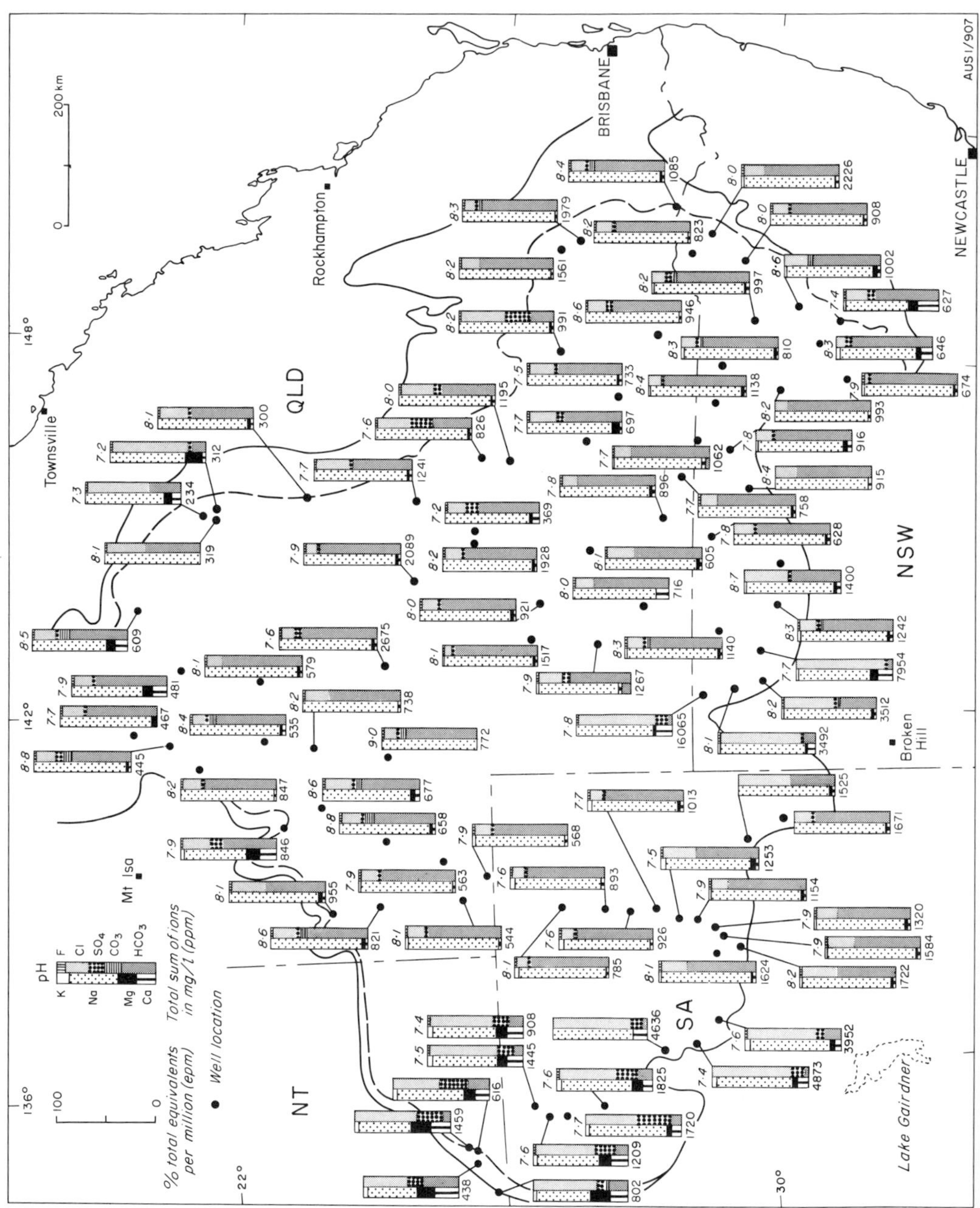

Fig. 5-18. Chemical compositions of ground water from selected flowing artesian wells tapping aquifers of Figure 5-17. (Habermehl, 1980, fig. 17, reproduced by courtesy of the Bureau of Mineral Resources, Canberra.)

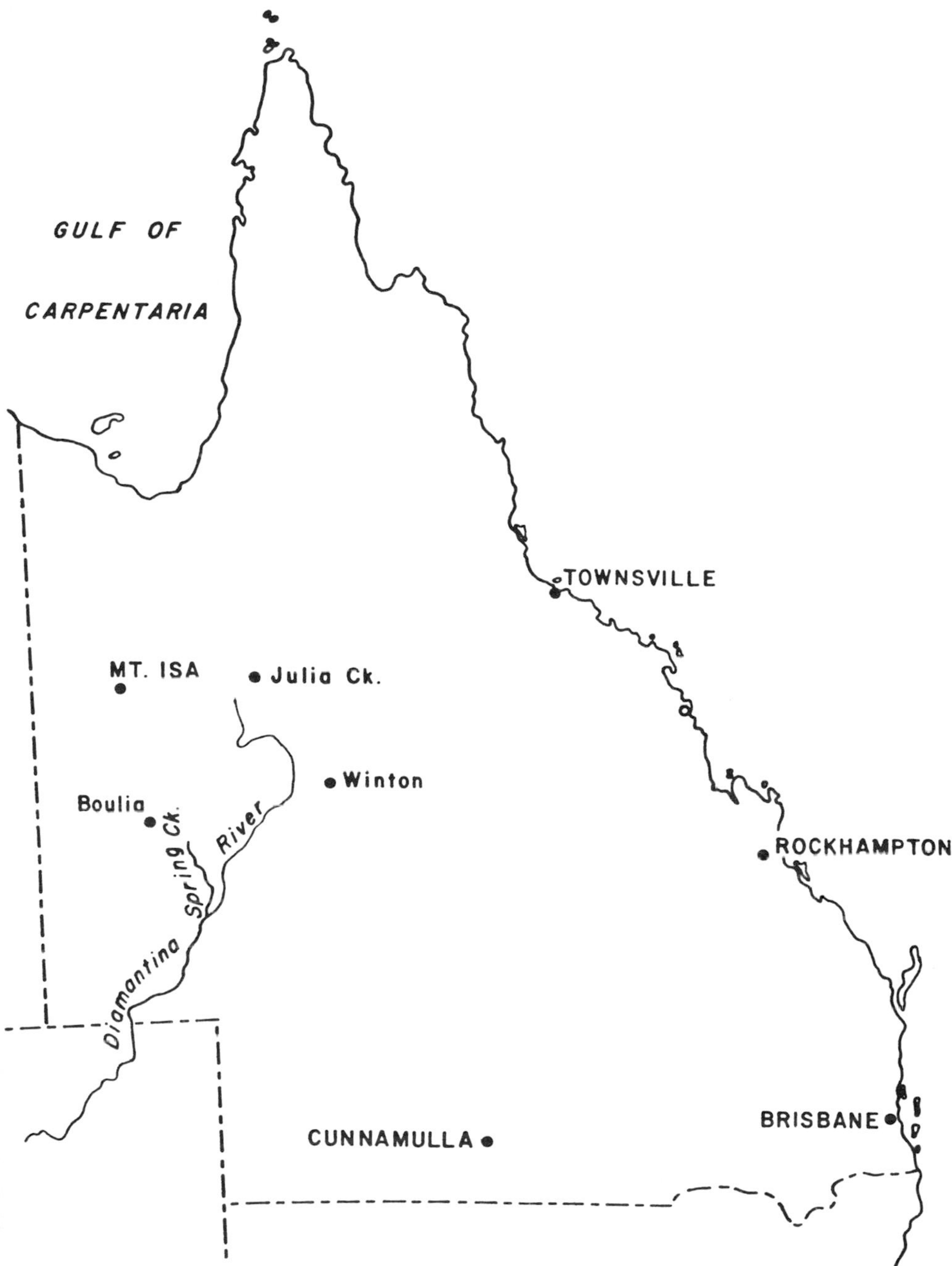

Fig. 5-19. Location map, Queensland portion of the Great Artesian Basin.

study of a large area of the eastern part of the Great Artesian Basin about Cunnamulla, concluded that the variations in the principal Jurassic aquifer reflect variations in the rate of infiltration of recycled salt throughout the late Quaternary.

We choose two aspects of the Great Artesian Basin to illustrate general principles: the age of the water, and the initial decline in production.

Habermehl (1980) quotes water velocities of 1 to 5 m per year (q/f), so the age of the water may be up to 100,000 years for every 100 km from the intake area. Water can be dated up to 50 years using tritium (^3H), an isotope of hydrogen with two neutrons in addition to the proton in its nucleus; and from 600 to about 30,000 years using carbon 14. Thus the age, or residence time, of the water must be estimated from hydraulic data for the time being (there is hope that chlorine 36, with a half-life of 308,000 years, will become a reliable tool for dating old water).

It is sometimes argued (and has been argued informally for the Great Artesian Basin) that old water indicates that it is a non-renewable resource that should be treated the same as mining any other non-renewable resource. There is not the slightest doubt that the water in the Great Artesian Basin is being renewed from the intake area – the potentiometric maps show that – and it could even be argued that production will improve the quality of the water with time.

The matter of high initial production and subsequent decline to a stable rate can be considered to have three components. The potentiometric surface on completion of the first well is at its natural level, and during the early period of production, this is drawn down to a level that depends on the hydraulic properties of the aquifer and its depth below the surface. The aquifer's energy is therefore initially high at the well-site, but declines as it adjusts to steady flow and the development of a stable hydraulic gradient to the well.

The other two components of this decline come under the heading of *specific storage*. Specific storage is the sum of the water stored in the aquifer by virtue of the compressibility of the solid framework of the aquifer, and water stored by virtue of its own compressibility. When the aquifer pressure is relieved by production, Terzaghi's relationship comes into play, and the decline in water pressure is compensated by an increase in the effective stress – which in turn compacts the rock a little, expelling a further quantity of water ($S = \sigma + p \simeq$ constant). The water also expands a little as a result of this relief of pressure. Production stabilizes when the aquifer stabilizes under the new conditions.

SELECTED BIBLIOGRAPHY

Airy, P.L., Calf, G.E., Campbell, B.L., Hartley, P.E., Roman, D., and Habermehl, M.A., 1979. Aspects of the isotope hydrology of the Great Artesian Basin, Australia. *Isotope Hydrology* 1978, 1: 205-219. (Proc. Internat. Symp. on Isotope Hydrology, International Atomic Energy Agency and UNESCO, Neuherberg, Federal Republic of Germany. I.A.E.A., Vienna.)

Badon Ghijben, W., 1889. See Drabbe, J., and Badon Ghijben, W., 1889.

Beckett, P., 1953. Qanats around Kirman. *J. Royal Central Asian Soc.*, 40: 47-58.

Breuninger, F.W., 1719. *Fons Danubii primus et naturalis oder die Urquelle des weltberühmten Donaustroms*. Tübingen, 388 pp.

Buchan, S., 1938. The water supply of the County of London from underground sources. *Memoir Geol. Surv. Great Britain*, 260 pp.

Butler, M.A., 1933. Irrigation in Persia by kanáts. *Civil Engineering*, 3 (2): 69-73.

Day, J.W.B., and Rodda, J.C., 1978. The effects of the 1975-76 drought on groundwater and aquifers. *Proc. Royal Society London*, A363: 55-68.

Dickey, P.A., and Cox, W.C., 1977. Oil and gas reservoirs with subnormal pressures. *Bull. American Ass. Petroleum Geologists*, 61 (12): 2134-2142.

Drabbe, J., and Badon Ghijben, W., 1889. Nota in verband met de voorgenomen putboring nabij Amsterdam. *Tijdschrift van het Koninklijke Instituut van Ingenieurs*, 1888-1889: 8-22.

Freeze, R.A., and Witherspoon, P.A., 1966. Theoretical analysis of regional groundwater flow. 1. Analytical and numerical solutions to the mathematical model. *Water Resources Research*, 2 (4): 641-656.

Freeze, R.A., and Witherspoon, P.A., 1967. Theoretical analysis of regional groundwater flow. 2. Effect of water-table configuration and subsurface permeability variation. *Water Resources Research*, 3 (2): 623-634.

Freeze, R.A., and Witherspoon, P.A., 1968. Theoretical analysis of regional groundwater flow. 3. Quantitative interpretations. *Water Resources Research*, 4 (3): 581-590.

Ghijben, W.B., 1889. See Drabbe, J., and Badon Ghijben, W., 1889.

Glover, R.E., 1959. The pattern of fresh-water flow in a coastal aquifer. *J. Geophysical Research*, 64 (4): 457-459.

G[orgy], S., and S[alah], M.M., 1974. Mediterranean Sea. *In: The New Encyclopaedia Britannica* (15th ed.), *Macropaedia*, Vol. 11, pp. 854-856. Encyclopaedia Britannica Inc., Chicago, London, etc.

Graves, W., 1975. Iran, desert miracle. *National Geographic*, 147 (1): 2-47.

Habermehl, M.A., 1980. The Great Artesian Basin, Australia. *Australia, Bureau of Mineral Resources J. Australian Geology and Geophysics*, 5: 9-38.

Halley, E., 1687. An estimate of the quantity of Vapour raised out of the Sea by the warmth of the Sun; derived from an Experiment shown before the Royal Society, at one of their late Meetings. *Philosophical Trans. Royal Society London*, 16 (189): 366-370.

Halley, E., 1691. An account of the circulation of the watry Vapours of the Sea, and of the cause of Springs. *Philosophical Trans. Royal Society London*, 17 (192): 468-473.

Herzberg, B., 1901. Die Wasserversorgung einiger Nordseebäder. *Journal für Gasbeleuchtung und verwandte Beleuchtungsarten sowie für Wasserversorgung*, 44: 815-819, 841-844.

Hill, G.A., Colburn, W.A., and Knight, J.W., 1961. Reducing oil-finding costs by use of hydrodynamic evaluations. *In: Economics of petroleum exploration, development, and property evaluation*. Prentice-Hall, Englewood Cliffs, N.J., pp. 38-69.

Hitchon, B., 1969a. Fluid flow in the western Canada sedimentary basin. 1. Effect of topography. *Water Resources Research*, 5 (1): 186-195.

Hitchon, B., 1969b. Fluid flow in the western Canada sedimentary basin. 2. Effect of geology. *Water Resources Research*, 5 (2): 460-469.

Hitchon, B., and Hays, J., 1971. Hydrodynamics and hydrocarbon occurrences Surat Basin, Queensland, Australia. *Water Resources Research*, 7 (3): 658-676.

Hubbert, M.K., 1940. The theory of ground-water motion. *J. Geology*, 48 (8): 785-944.

Ineson, J., and Downing, R.A., 1964. The ground-water component of river discharge and its relationship to hydrology. *J. Instn Water Engineers*, 18 (7): 519-541.

Ineson, J., and Downing, R.A., 1965. Some hydrogeological factors in permeable catchment studies. *J. Instn Water Engineers*, 19 (1): 59-80.

Käss, W., 1969. Schrifttum zur Versickerung der oberen Donau zwischen Immendingen und Fridingen (Südwestdeutschland). *Steirische Beiträge zur Hydrogeologie*, 21: 215-246.

Käss, W., 1972. Die Versickerung der Oberen Donau, ihre Erforschung und die Versuche 1969. *Geologisches Jahrbuch*, C2: 13-18.

Käss, W., and Hötzl, H., 1973. Weitere Untersuchungen im Raum Donauversickerung-Aachquelle (Baden-Württemberg). *Steirische Beiträge zur Hydrogeologie*, 25: 103-116.

King, F.H., 1892. Observations and experiments on the fluctuations in the level and rate of movement of ground-water on the Wisconsin Agricultural Experiment Station Farm and at Whitewater, Wisconsin. *U.S. Dept. Agriculture, Weather Bureau, Bulletin* no. 5 (75 pp.).

Kraus, L., 1969. Erdöl- und Erdgaslagerstätten im ostbayerischen Molassebecken. *Erdoel-Erdgas-Zeitschrift*, 85: 442-454.

Lemcke, K., 1976. Übertiefe Grundwässer im süddeutschen Alpenvorland. *Bull. Vereinigung Schweizerischer Petroleum-Geologen und -Ingenieure*, 42 (103): 9-18.

Lemcke, K., and Tunn, W., 1956. Tiefenwasser in der süddeutschen Molasse und in ihrer verkarsteten Malmunterlage. *Bull Vereinigung Schweizerischer Petroleum-Geologen und -Ingenieure*, 23 (64): 35-56.

Lyell, C., 1867. *Principles of Geology or the modern changes of the Earth and its inhabitants* (10th ed.), *vol.* 1. John Murray, London, 671 pp.

Mariotte, E., 1717. Traité du mouvement des eaux et des autres corps fluides. *In: Oeuvres de Mr. Mariotte*, Pierre van der Aa, Leiden, pp. 321-476.

Noel, E., 1944. Qanats. *J. Royal Central Asian Soc.*, 31: 191-202.

Norton, W.H., 1897. Artesian wells of Iowa. *Iowa Geol. Survey*, 6: 113-428.

Ogilvie, C., 1955. The hydrology of the Queensland portion of the Great Australian Basin. *In*: Queensland, Artesian Water Investigation Committee, *Artesian water supplies in Queensland*. Government Printer, Brisbane, Appendix H.

Ohrt, F., 1947. Water development and salt water intrusion on Pacific islands. *J. American Water Works Ass.*, 39 (10): 979-988.

Perrault, P., 1674. *De l'origine des fontaines*. Pierre le Petit, Paris, 353 pp. (Not sighted.)

Perrault, P., 1967. *On the origin of springs*. (English translation by A. LaRocque.) Hafner, New York and London, 209 pp.

Queensland, Artesian Water Investigation Committee, 1955. *Artesian water supplies in Queensland*. Government Printer, Brisbane, 79 pp + 2 appendices.

Sandford, K.S., 1935. Sources of water in the north-western Sudan. *Geographical Journal*, 85 (5): 412-431.

Schröckenfuchs, G., 1975. Hydrogeologie, Geochemie und Hydrodynamik der Formationswässer des Raumes Matzen-Schönkirchen Tief. Erdoel-Erdgas-Zeitschrift, 91: 299-321.

Smith, A., 1953. *Blind white fish in Persia*. Allen & Unwin, London, 231 pp.

Spain, Presidencia del Gobierno, 1964. *Canarias: anexo al Plan de Desarrollo Económico y Social* 1964-1967. Edición del Boletin Oficial del Estado, Madrid, 518 pp.

Thesiger, W., 1959. *Arabian sands*. Longmans, London, 326 pp.

Tóth, J., 1962. A theory of groundwater motion in small drainage basins in central Alberta, Canada. *J. Geophysical Research*, 67 (11): 4375-4387.

Tóth, J., 1963. A theoretical analysis of ground-water flow in small drainage basins. *J. Geophysical Research*, 68 (16): 4795-4812.

Tóth, J., 1979. Patterns of dynamic pressure increment of formation-fluid flow in large drainage basins, exemplified by the Red Earth region, Alberta, Canada. *Bull. Canadian Petroleum Geology*, 27 (1): 63-86.

Van Everdingen, R.O., 1968. Studies of formation waters in western Canada: geochemistry and hydrodynamics. *Canadian J. Earth Science*, 5: 523-543.

Veatch, A.C., 1906. Fluctuation of the water level in wells, with special reference to Long Island, New York. *U.S. Geol. Surv. Water-Supply and Irrigation Paper* 155 (83 pp.).

Versluys, J., 1917. De beweging van het grondwater. *Water*, 1: 23-25, 44-46, 74-76, 95.

Versluys, J., 1919. De duinwater-theorie. *Water*, 3 (5): 47-51.

Versluys, J., 1920. The theory of dune water. *Water Services*, 22: 182-184. (Incomplete translation into English of Versluys, 1919, without figures.)

Walpole, G.F., 1932. An ancient subterranean aqueduct west of Matruh. *Survey of Egypt Paper* (*Egypt, Maslahat al-Misahah*), 42 (40 pp.).

Wentworth, C.K., 1952. The process and progress of salt water encroachment. *Union Géodésique et Géophysique Internationale, Association Internationale d'Hydrologie Scientifique, Assemblée Générale de Bruxelles* 1951, 2: 238-248.

Whitaker, W., and Tresh, J.C., 1916. The water supply of Essex from underground sources. *Memoir Geol. Surv. Great Britain*, 510 pp.

Whitehouse, F.W., 1955. The geology of the Queensland portion of the Great Artesian Basin. *In*: Queensland, Artesian Water Investigation Committee, *Artesian water supplies in Queensland*. Government Printer, Brisbane, Appendix G.

Wulff, H.E., 1966. *The traditional crafts of Persia: their development, technology, and influence on eastern and western civilizations*. M.I.T. Press, Cambridge, Massachusetts, 404 pp.

Wulff, H.E., 1968. The qanats of Iran. *Scientific American*, 218 (4): 94-105.

6. MOVEMENT OF PORE WATER, AND ABNORMALLY HIGH PORE PRESSURES

The distinction between ground water and pore water may not be very logical, but it has the merit of distinguishing the readily-exploitable fresher pore water near the surface from the brackish to salty water in the pore spaces of most sedimentary rocks at greater depth. As always, such distinctions recognize tendencies only, for there are areas (such as the Niger delta; see Dailly, 1976, p. 96, fig. 3) where fresh water is found in sands to depths of two or three kilometres. The distinction is also seen as a distinction between *meteoric* water and what is called *connate*, the former being derived 'recently' from rainfall, the latter being defined as water that was trapped in the pore spaces when the sediment accumulated.

The difficulty with the definition of connate water is that we cannot accept that significant amounts of water were *trapped* in the pores when the sediment accumulated. If one considers the bulk water in the pores of a rock unit, then it is an acceptable interpretation of the pore water of muds and mudstones (although the chemistry of the water may well have changed) but it is not acceptable for the pore water in porous and permeable sediments such as sands and some limestones. The doctrine of uniformitarianism suggests that some sediment masses with 'connate' water passed through a ground-water stage earlier in their history. There is little doubt that pore water in sedimentary rocks is normally in motion, and that it cannot therefore be strictly 'connate' – and this applies also to the pore water of most mudstones. An important mechanism for this motion is the compaction of the sediments.

COMPACTION OF SEDIMENT

In its simplest terms, compaction is a diagenetic process by which sediment under the weight of its own overburden is compressed so that its bulk volume is reduced and its bulk density increased. Compaction can only be achieved by the reduction of pore volume, so compaction leads to the migration of the pore fluids towards positions of lower energy. The loss of porosity leads also to loss of permeability (as we have seen in Chapter 3); and the concomitant increase in bulk density leads to a more competent rock, but a thinner one.

While we shall concentrate on the mechanical aspects of compaction, it must be remembered that many other influences, related and unrelated, may be present. Clearly, increasing the temperature of the rock during burial reduces the effective viscosity of the solid constituents as well as the liquid, and time increases the strain in the rock. Chemical changes alter both solids and liquids, leading to what some call consolidation rather than compaction (but the nomenclature is confused).

Compaction of sands

Sand on the sea floor is moved and agitated by the waves and swell of the sea, by currents, and by seismic shocks. This leads to the sorting of the sediments, and to the arrangement of the grains in a more or less stable packing. These influences extend to the edges of the continental shelves (the influence of swell, not waves, can be felt at such depths). But we must remember that the sediment on the sea-floor is not normally accumulating into the stratigraphic record as such: it is only that part of the sediment that reaches a position in which the energy of the environment is insufficient to move it further – that part that comes to be permanently below baselevel – that accumulates into the stratigraphic record. This is the process of winnowing, and the process behind facies changes. While it is a process that cannot be observed because the human life-span is too short, we may assume that a sediment passes into the stratigraphic record no worse sorted than it was before. (See Chapman, 1973 or 1976, pp. 1-18, for a more detailed discussion of this difficult point.)

Studies of sand bodies have revealed that variation of porosity due to facies is far greater than variation due to compaction with depth; but that sands appear to have a linear compaction trend with depth (Maxwell, 1964). As the load increases with depth, so the effective stress increases and porosity is reduced. The grains deform elastically at first, but, due to the time dimension, recovery is not complete if the load is removed long afterwards. (See Stephenson, 1977, for an interesting discussion of the maximum effect of any one component, in particular, temperature.)

The loss of porosity is relatively small in sand compaction, and the loss of permeability (provided there is no chemical precipitate or cement) is also relatively slight. Taking representative compaction figures from Maxwell (1964) and evaluating the combined porosit term from p. 60, $f^{3.5}/(1-f)^2$, we estimate that reduction of porosity from 35% to 25% results in a 75% loss of permeability to 25% of its former value. While this is a large proportional reduction, the permeability at 25% porosity may still be large.

Under severe loading, the grains may fracture and change shape: quartz overgrowths and cements also change pore geometry. Fracture of the grains reduces permeability to a much greater extent through the increase in specific surface as well as the greater loss of porosity due to changed geometry.

Compaction of muds, clays, and 'shales'

The compaction of muds is a complex process that has both physical and chemical components. We shall follow previous practice in regarding the mechanical processes as dominant, and then consider the modifications due to other causes. In many applied geological contexts, the term shale is used to embrace all fine-grained compactible sediments other than carbonates.

Muds and mudstones contain a significant, but not necessarily dominant proportion of minerals that deform plastically under load, that is, material that does not recover its former dimensions when the load is removed. Many of them are platy in form, so that contacts between grains tend to be surfaces rather than points of contact. The most significant consequence of these properties from our point of view is that compaction of mudstones not only reduces porosity permanently, but also seriously and permanently reduces permeability on account of the restriction of the passages from one pore to another. This effect has important consequences, because the loss of permeability also inhibits the expulsion of the fraction of the pore water that must escape if further compaction is to take place. We shall pursue this matter later.

There are two practical approaches to the determination of the relationship between porosity – and bulk density – and depth: direct measurement, and indirect measurement by geophysical means. Direct measurement, which requires very careful work, was carried out by Hedberg (1926, 1936) on Tertiary shales in boreholes in Venezuela; and by Athy (1930) on upper Palaeozoic shales in Oklahoma (Fig. 6-1). Hedberg was careful to point out that the direct relationship is between pressure and porosity, not depth and porosity; but he recognized the value and need for depth-porosity relationships.

Hedberg (1936) came to a number of important conclusions:

a) there are four broad stages of compaction: first, a mechanical re-arrangement of the particles; then a de-watering stage with expulsion of some of the pore water and adsorbed water; then mechanical deformation of the grains; and finally, a recrystallization stage. These stages overlap.

b) porosity is a better indicator of compaction than bulk density; and that reduction of porosity is related in the early stages to the overburden pressure. With overburden pressures from 0 to 800 psi (to 5.5×10^6 N m^{-2}) he found an exponential relationship

$$P = 67.214\, G^{-0.1047},$$

where P is the porosity (%) and G is the overburden pressure in psi (estimated from his measurements). From 800 psi to 6000 psi (to 4.1×10^7 N m^{-2}) a linear relationship

$$P = 34.86 - 0.00421\, G.$$

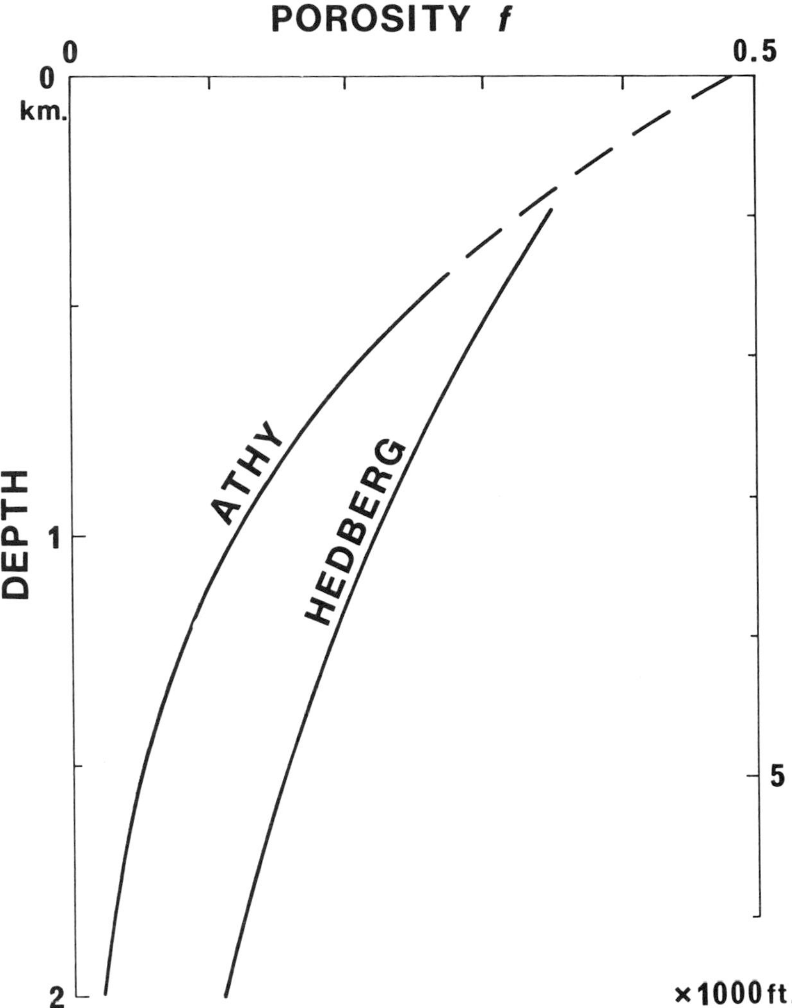

Fig. 6-1. Shale compaction curves of Athy (1930) and Hedberg (1936), the latter generalized.

From 6000 psi to 10000 psi (to 6.9×10^7 N m^{-2}), he found a different linear relationship

$$P = 13.93 - 0.0006935\, G\,.$$

c) he accepted that there is a need for a single function relating porosity to pressure, and found that for depths greater than those at which the overburden pressure is greater than 200-300 psi,

$$P = 40.22\,(0.9998)^G. \tag{6.1}$$

d) he found that depth in feet is *approximately* equal numerically to the overburden pressure in psi (Hedberg, 1936, pp. 259-260).

Athy (1930) studied upper Palaeozoic (Permian to base Pennsylvanian) shales in Oklahoma, and proposed the relationship

$$f = f_0 e^{-cz}, \tag{6.2}$$

where f is the porosity at depth z, f_0 the porosity depth $z = 0$, and c, a coefficient (with a value 1.42×10^{-3} m^{-1} for his data). He had to extrapolate the top 1400 ft (430 m) of his curve, and some doubt has been cast on his results due to possible tectonic forces having contributed to the compaction.

Athy's curve has been widely used by subsequent workers (e.g., Rubey and Hubbert, 1959; Chapman, 1973; Magara, 1978), and attempts at reconciling Athy's curve with Hedberg's (e.g., Weller, 1959) have not been convincing. It has generally been considered that Hedberg's curve for Tertiary shales would approach Athy's curve for Palaeozoic shales with time.

It is important to realize that neither curve implies the history of porosity reduction with time and burial, but only the relationship between porosity and depth or pressure now. Therein lies the obvious weakness of such empirical formulae, for temperature and time do not appear explicitly in either. And one must always remember that compaction is not always only a vertical process: there may be lateral spreading also in some situations (such as deltas).

Recently, Chapman and Fitz-Gerald (in prep.) took a different approach. If rocks in the context of geological time behave not as solids but as liquids, then compaction can be modelled by regarding rocks as compressible liquids. They propose the dimensionless expression (still without time or temperature)

$$f^* = \frac{\gamma}{1 + (\gamma - 1)\, e^{\gamma z^*}}, \tag{6.3}$$

where $f^* = f/f_0, \gamma = \rho_s/(\rho_s - \rho_w) f_0$ and $z^* = z/a$, where a is a scale length. In theory, the scale length can be measured from experimental data: it is the length that makes $d\rho/d\sigma$ unity for small values of σ. Figure 6-2 shows f^* plotted against z^* for γ-values of 3, 3.5, and 4, within which range most shale sections will fall.

While the dimensional plots of Hedberg's (1936) and Athy's (1930) data seemed conflicting, the dimensionless plots agree well with the field data, the values of γ being 3.9 and 3.4 respectively, when a scale length of 2,000 m is assigned to Athy's data, and 5,000 m to Hedberg's. The scale length is related to the mechanical properties of the sediment by definition, but it may also take other influences, such a temperature and time, into account.

The same model applies to the mechanical compaction of sand (or any other geological material) as well as shale, but the scale length for sand is larger, and so the dimensional plot is virtually a straight line.

These results have not yet been widely tested, and so should not be accepted uncritically. Much work remains to be done on sediment compaction trends with

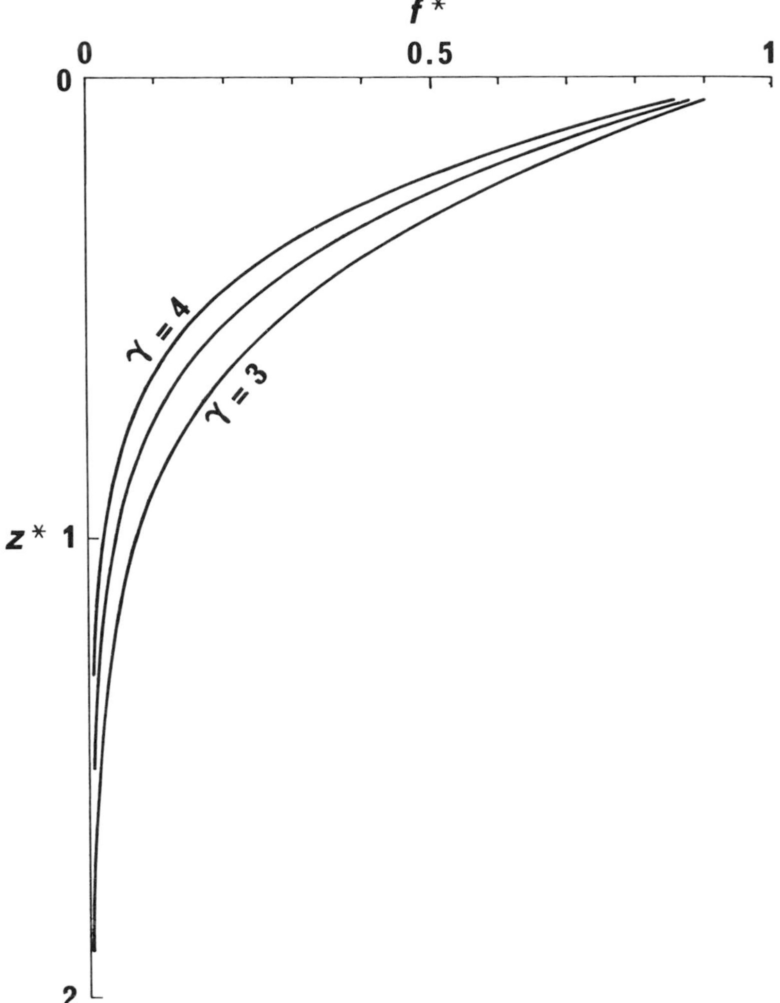

Fig. 6-2 Dimensionless compaction curves of Chapman and Fitz-Gerald (in preparation). $f^* = f/f_0$, $\gamma = \rho_s/(\rho_s - \rho_w)f_0$ and $z^* = z/a$ where a is a scale length.

depth, and the role of chemical diagenesis in these.

From a practical point of view, however, we can rarely obtain the data we need by direct measurement, and must resort to indirect, geophysical, methods in boreholes if we are to obtain the local compaction trend of shales with depth. For this, the most reliable method is to use the *sonic log*, and to plot the logarithm of the sonic transit time in shale (Δt_{sh}) against depth. A great mass of data from various parts of the world indicates that over the normally-compacted part of the sedimentary sequence, these points plot on a straight line (with some scatter). Figure 6-3 shows

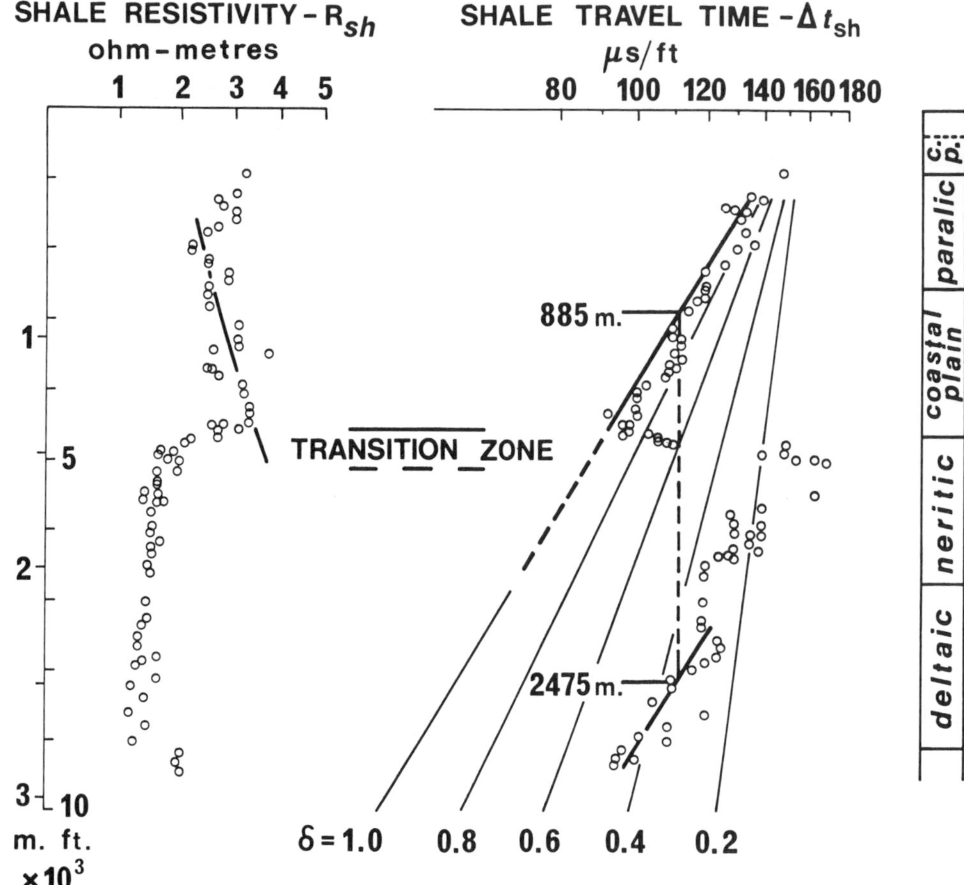

Fig. 6.3. Shale resistivities and shale travel times in a well in Borneo (courtesy of the Royal Dutch/Shell Group).

shale resistivity and shale transit time plotted on a logarithmic scale against depth for a well in Tertiary sediments in Borneo. Such plots of shale transit time can be expressed by an empirical formula of the same form as Athy's for compaction,

$$\Delta t_{sh} = \Delta t_0 \, e^{-bz}, \tag{6.4}$$

where Δt_0 is the extrapolated value of Δt_{sh} at depth $z = 0$, and b $[L^{-1}]$ is the slope $(\log \Delta t)/z$. It is found that Δt_0 is commonly about 540 μs m^{-1} (165 μs/ft) corresponding to a sonic velocity in shales of about 1,850 m s^{-1} (6,000 ft/sec) and f_0 of about 0.5; but the slope varies from area to area.

It must be noted that the common assumption that porosity is proportional to shale transit time is only true if the slope b is the same as the slope c in Athy's equation. Very little data has been published, but it seems to the author that this is only true for very thick shale sections with little sand, and that the slope of the shale compaction line, c, can be larger than b by a factor of at least 3 when there is a significant number of sands in the section. (The slope c can be estimated from the Formation Density Log if one assumes that the bulk density is inversely proportional to porosity.)

Confidence in the use of the sonic log comes from the fact that when shales with abnormally high pore pressures are encountered, the shale transit time *increases* with depth (velocity decreases) in what is known as the transition zone (Fig. 6-3); and this departure from the normal trend usually agrees with drilling data very closely (more closely than that of the Formation Density Log, for reasons that are not understood). Shales that show an increase in transit time with depth into a zone of abnormally-high pores pressures are inferred to be undercompacted for their depth – they have not compacted fully because the commensature part of the pore water has not been expelled. We shall look, therefore, at pore pressures in stratigraphic sequences.

PORE PRESSURES

A very large number of water wells, and boreholes drilled for petroleum, provide evidence that the pore pressures in the upper few kilometres of sedimentary basins are normal hydrostatic: they will support a column of the water to near the land surface, or to sea-level in low-lying areas (Fig. 6-4). These observations indicate that the water in the pores at depth is in physical continuity to the surface, and we may regard the sediment as having accumulated in water with as much logic as the more usual statement that the rocks contain water. If we wish to call this water 'connate', then we must redefine the term because this water is not *trapped* in the sediment.

Pore pressures may be approximated by the relationship found in equation (1.2):

$$p = \rho g z, \tag{6.5}$$

where ρ is the mean mass density of the water, and z the depth at which the pressure is required to be known. This is an approximation only: the density may vary with depth due to varying salinity, its compressibility, and its tendency to expand with depth due to increasing temperatures. Furthermore, the relationship implies that the free water surface coincides with that at which $z = 0$. This is usually sufficiently nearly the case, but there are areas, as we have seen in Chapter 5, in which the free water surface is not near the level where $z = 0$. On the one hand, there are artesian aquifers: on the other, aquifers with abnormally low pressures (such as those in the Molasse basin of southern Germany).

There are also data obtained from many parts of the world from petroleum exploration drilling that indicate abnormally high pore pressures, in some cases approaching the pressure exerted by the total overburden:

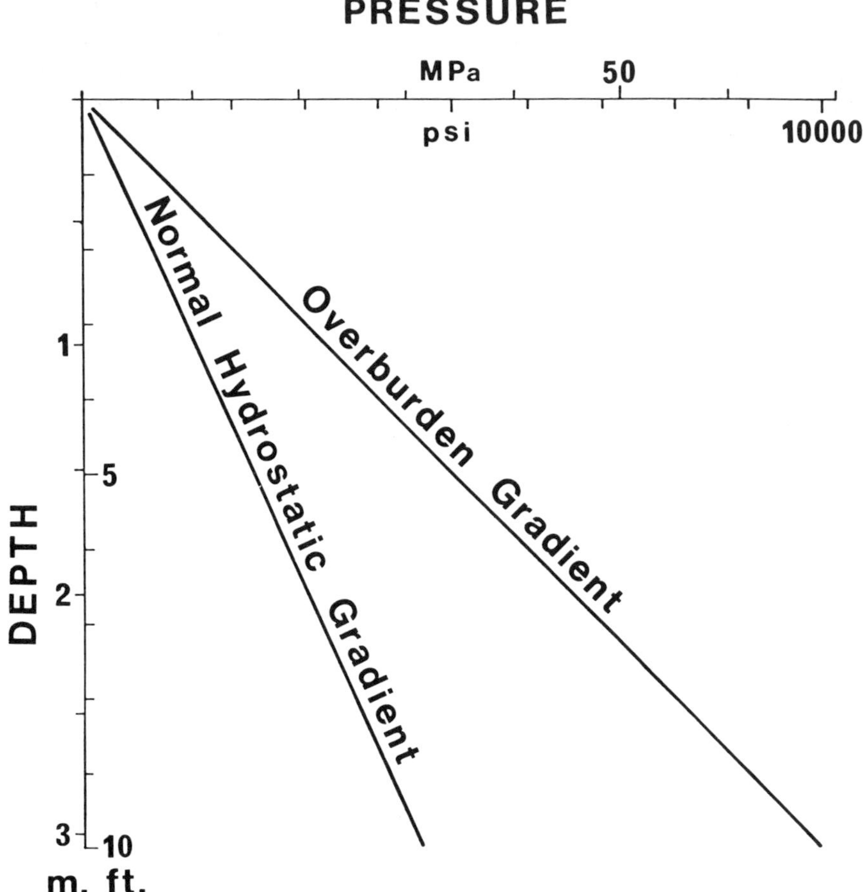

Fig. 6-4. Typical pressure-depth plot of sedimentary sequence.

$$p \rightarrow \rho_{\mathrm{bw}} g z, \tag{6.6}$$

where ρ_{bw} is the mean bulk wet density of the overburden.

This relationship is also an approximation because ρ_{bw} may very – and usually does – not only with depth but also with lithology. The value of ρ_{bw} usually falls between 2,000 kg m^{-3} for unconsolidated sediments of high porosity and 2,400 kg m^{-3} (2.0 g/cm^3 to 2.4 g/cm^3). Hence the overburden pressure gradient is commonly about 22,000 N m^{-2} per metre (0.22 kg/cm^2 per metre). These pressures are known as *geostatic* pressures, and their gradients as geostatic gradients. *Lithostatic* is a common but undesirable synonym: geostatic has priority (Dickinson, 1951; 1953, p. 425). We shall, however, use the word *overburden* because it has the merit not only of priority but also of implying that the total load of solids and fluids is involved.

Thus, over the normally-pressured part of the sequence, we may regard the overburden pressure as the total normal stress on a horizontal plane at depth z. We have seen in Chapter 1 that Terzaghi (1936), in the context of soil mechanics, postulated that the total normal stress on such a surface consists of two partial pressures: the pressure or stress transmitted through the solid framework, from grain to grain, and the pressure exerted by the pore water,

$$S = \sigma + p. \text{ (Terzaghi's relationship)} \tag{6.7}$$

The stress transmitted through the solid framework, σ, which Terzaghi called the *effective* stress, is the stress that compacts a sedimentary rock. Terzaghi called the pore pressure the *neutral* stress because it does not directly deform the grains.

It will be recalled that the field of fluid pressures is inferred to exist throughout the space occupied by the fluid *and* the contained solids, and that this force supports part of the weight of the solids. Hence the effective stress may be seen as the difference between the overburden pressure and the pore pressure:

$$\sigma = S - p. \tag{6.7a}$$

This is shown graphically in Figure 6-5.

The question now arises: what is the nature of abnormally-high pore pressures, such as are found in so many parts of the world?

Dickinson (1951, 1953) showed in his classic paper that stratigraphy, not depth, determined the presence of abnormally-pressured reservoirs in the Louisiana Gulf Coast. The top of abnormal pressures occurs beneath the dominantly sandy se-

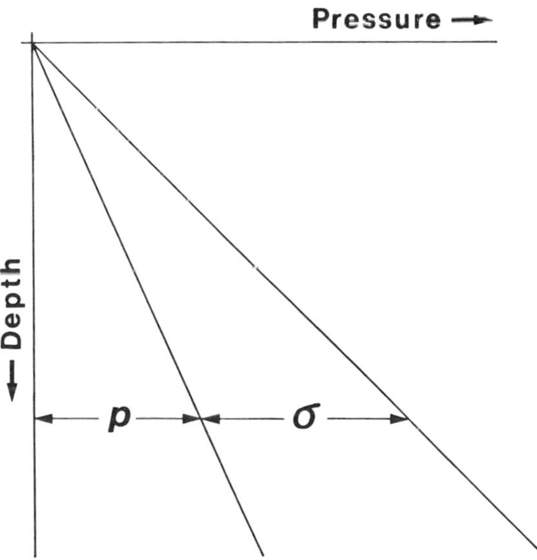

Fig. 6-5. Effective stress is the difference between overburden pressure and pore-water pressure (here, normal hydrostatic).

quence, in sediments of progressively younger age from north to south. The stratigraphic context of shales with abnormally-high pressures is overwhelmingly that of a dominantly regressive sequence, in which paralic and neritic sands overlie neritic shales (Chapman, 1973, 1976, 1977). The sand content of the sequence (in terms of beds of permeable sand) is significant because abnormal pressures are not normally found in regressive sequences where the percentage of sand is large. Harkins and Baugher (1969) found 5% to 10% sand to be a regional indicator of impending high pressures for drilling in the U.S. Gulf Coast region.

The pressure-depth plot of a typical well in a typical thick Tertiary regressive sequence is shown in Figure 6-6a. Below the top of what is known as the transition zone, pore pressures increase above hydrostatic on a gradient ($\Delta p/\Delta z$) typically much greater than the overburden pressure gradient; but before reaching the overburden pressure, the gradient decreases and the maximum value of pore pressure is commonly found to be about 90% of the overburden pressure. The ratio of pore pressure to overburden pressure at a given depth, p/S, is given the symbol λ (Hubbert and Rubey, 1959, p. 142): it is regarded as the proportion of the total overburden supported by the fluid pressure. So, when $p = S$, $\lambda = 1$, and the total overburden is supported by the pore fluid.

Reverting to Terzaghi's relationship, Equation 6.7, and substituting $p = \lambda S$, we can write

$$\sigma = S(1 - \lambda)\,. \tag{6.7b}$$

As the pore-fluid pressure, increases towards the overburden pressure, $\lambda \rightarrow 1$ and the effective stress that compacts the rock approaches zero. Boreholes that penetrate abnormally-pressured shales below a normally-pressured sand/shale sequence usually reveal that the shales are undercompacted for their depth, having greater porosity that would be normal for their depth.

Undercompaction is indicated by the reduction of electrical resistivity and reduction of sonic velocity, as measured in the wall of the borehole by borehole logging devices. Figure 6-3 shows the resistivities and sonic travel times measured in shales in a borehole in Borneo. It will be seen that both depart from the normal trend at about the same depth, and the departure is consistent with an *increase* in porosity with increasing depth in the transition zone.

There is also some direct evidence of these features acquired during the drilling of such sequences: the increase in drilling rate when abnormal pressures are reached (known as a 'drilling break'), and the occurrence of 'heaving shales'. The rate of drilling depends on a number of factors under the control of the driller (mud weight, weight on bit, rotation speed, hydraulic energy of the mud circulation, type of bit), and some not under his control (the type of rock and its drillability, the pore fluid pressure). If those factors under the control of the driller are kept constant, then the penetration rate is found to be a fair guide to the lithologies being drilled. Sand, for example, drills more easily than shale. But it was found that as the bit entered the

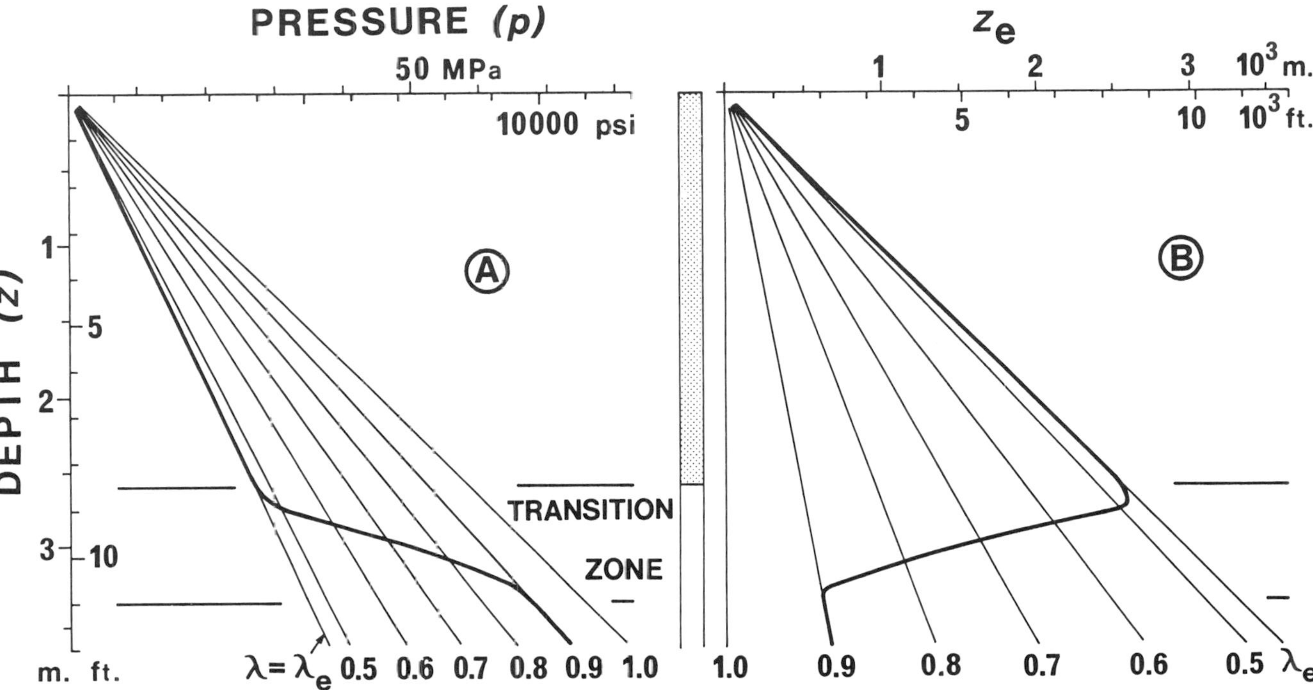

Fig. 6-6. Pressure-depth plot of typical young regressive sequence (A), and plot of equilibrium depth against true depth (B).

transition zone to abnormally-high pressures, the penetration rate increased remarkably (in many parts of the world this became the standard method of detecting the top of abnormal pressures while drilling). The cause of this increase is seen as the reduction or reversal of the fluid potential gradient at the bottom of the hole, from being downwards to being upwards (or reduced downwards); and to some extent, the increase in porosity, and decrease in cohesive strength of the shale.

It is evident that the pore water in the transition zone is at a much higher *potential*, as well as pressure, than that in the overlying normally-pressured sequence, and that consequently there is a tendency for the pore water to flow down this potential gradient (which we may take to be vertical if the cause of the abnormality is gravitational). As the pore water flows down this gradient, so the effective stress increases in the direction of flow (upwards), and the shale is more compact in this direction. This loss of porosity in the direction of flow also involves a loss of permeability. Hence the gradient of pore pressure in the transition zone is seen as a potential gradient due to the flow of pore water. It is therefore independent of the overburden pressure gradient, and assumes a gradient that is a function of permeability – the lower the permeability of the shale, as a whole, in the transition zone, the steeper the gradient $\Delta p/\Delta z$.

The corollary is that much of the pore water in abnormally-pressured shales is original pore water (for which we can reasonably use the term *connate*), and that undercompaction is, as we have inferred, largely due to the failure of the pore water to move in sufficient quantities in response to the loading. This inference is supported by petroleum exploration experience: commercial accumulations of petroleum are rarely found where the pore pressures are abnormally high (see Fertl, 1976, pp. 300-311), but are common above the transition zone.

The mechanics of sediment compaction leads us to further inferences concerning the origin of abnormally high pore pressures. There is little doubt that on the geological time scale, the constituents of shales behave plastically, and that compaction of shales is irreversible beyond the extent that can be attributed to thermal expansion of the pore water and the solid framework, or to possible expansion when interlayer water is released to the pore spaces (these possible effects will be discussed later). Hence, we cannot accept that an undercompacted shale has ever been significantly more compacted – that the porosity of an undercompacted shale has never been much less than it is now.

The interdependence of pore pressures and effective stress is developed further by arguing that the degree of compaction of a shale that is abnormally pressured corresponds to the normal compaction of an identical sediment at a much shallower depth (Figs. 6-7 and 6-6b). Rubey and Hubbert (1959, p. 174) called this shallower depth the *equilibrium depth*, or *effective compaction depth*. Using the same symbols as before, but adding the suffix 'e' to denote their value at the equilibrium depth, Terzaghi's relationship can be written

$$\sigma = S_e(1 - \lambda_e)\,. \tag{y.7c}$$

Equating equations (6.7b) and (6.7c), and substituting $S = \rho_{bw}\, g\, z$, and $S_e = \rho_{bw}\, g\, z_e$, we obtain

$$(1 - \lambda)\rho_{bw}\, g\, z = (1 - \lambda_e)\rho_{bw}\, g\, z_e$$

and so to the approximation

$$z_e = \frac{(1 - \lambda)}{(1 - \lambda_e)} z,$$

which we write as an equality because the mean bulk wet densities above the two depths will not differ by much. The effective compaction depth is related to the actual depth by a factor that takes the pore-fluid pressures into account. Denoting this factor by δ (Chapman, 1972a),

$$\delta = \frac{z_e}{z} = \frac{(1 - \lambda)}{(1 - \lambda_e)}\,. \tag{6.8}$$

Strictly, λ_e is the proportion of the overburden supported by the *ambient* fluid (see Chapman, 1979), sea water if submarine, air if subaerial; but almost all known sediments with such abnormally-high pore pressures are submarine, so we may take normal hydrostatic water pressures ($p_e = \rho_w\, g\, z$) to define λ_e ($= p_e/S$) for the time being.

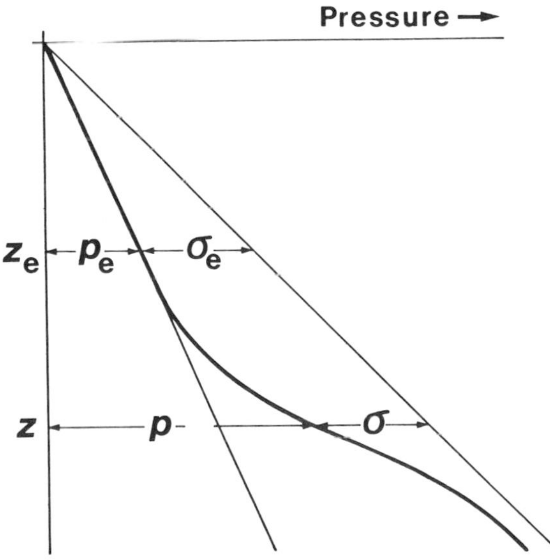

Fig. 6-7. Equilibrium depth (z_e) is the depth at which a normally-compacted mudstone would have the same value of effective stress as the overpressured mudstone at depth z.

The parameter δ is non-dimensional, and usually takes values between 0 (when $\lambda = 1$) and 1 (when $\lambda = \lambda_e$). In the present context of submarine abnormal pressures, δ may be taken as a non-linear measure of the degree to which the pore fluids have achieved mechanical equilibrium with the ambient fluid, or as a measure of the expulsion of pore fluids since the sediment accumulated. Equation (6.8) indicates that if the effective compaction depth is known or can be estimated, then the pore pressure can also be estimated.

It is also important to note that the effective compaction depth is an estimate of the maximum overburden at the time compaction equilibrium was lost, that is, the depth at which abnormal pressures began to develop.

As a practical example of the use of these concepts we can take Figure 6-3 (Sonic data). We assume that the straight-line part of the sequence is normally pressured, and that the value of the parameter δ is 1. Using the ratio z_e/z, we can construct lines of equal value of δ, the normal trend being the line $\delta = 1$, and the vertical from Δt_0 being $\delta = 0$. The smallest value of δ, the greatest abnormality, is about 0.2 (the smaller points are probably in borehole mud only, where the shale has washed out). This suggests that the equilibrium depth of the shale at the bottom of the transition zone is $0.2 \times 1{,}600 = 320$ metres, and that the abnormality began to develop soon after the accumulation of the more permeable sediments began. The pressure range estimated from mud-weights at a depth of 2,475 m was between 4.28×10^7 N m^{-2} and 4.66×10^7 N m^{-2} (436 and 475 kg/cm^2). At this depth, the value of δ is about 0.36 (885 divided by 2475). Since λ_e is approximately 0.46, and the overburden pressure gradient is about 2.26×10^7 N m^{-2} per metre, the pressure estimated from the sonic log would be about 4.52×10^7 N m^{-2} (460 kg/cm^2).

Now, the normal slope can also be taken, for practical purposes, as the slope corresponding to zero fluid potential gradient - the pore fluids being close to hydrostatic equilibrium. It follows that in sections in which the gradient of change of $\Delta t/\Delta z$ is smaller than the normal (more nearly vertical on the plot) the pressures are increasing with depth more rapidly that the hydrostatic, and the fluid potential gradient is upwards. Conversely, when the gradient of change is less than the normal, the fluid potential gradient is downwards. In Figure 6-3, there is evidence of both upwards and downwards migration of pore water within the main shale section penetrated, with a zone of separation a little deeper than 2,000 m. The environmental interpretation indicates a shallower environment for the sediments below this depth: this is a transgressive/regressive cycle, and there are probably more permeable rock units at greater depth, with pore fluids at lower potential. These trends are mirrored in the shale resistivity plot.

On another scale, we may apply the principles to the migration of pore water induced by compaction of any alternating sequence of campactible, relatively impermeable, beds and permeable, relatively uncompactible, beds – such as the alternating shales and sands in ,the neritic and paralic part of young regressive sequences.

In terms of the pore water, then, compaction of a water-saturated sediment creates differences of fluid potential that tend to drive the water in the direction of lower energy. The rate at which this movement can take place depends on the permeability of the sedimentary rock that is compacting *and* the resistance to all the water movement that results from the displacement of water from the compacting rock.

Thus a very permeable sand lens that is entirely surrounded by a relatively impermeable shale will only be able to compact to the extent that the pore water can escape from the total system.

In an alternating sequence of shales and sands compacting under gravity, the more compactible shales acquire a higher pore-fluid potential than the sands, and the pore fluid tends to migrate both upwards *and downwards* into normally-pressured adjacent sands (Fig. 6-8), and this movement will continue until compaction equilibrium has been reached – which, if the sequence is subsiding, may not be for many millions of years.

The commonly-held view that shale pore water migrates along the bedding 'because that is the direction of best permeability' is not in accord with what we know of the mechanics of pore water movement. Shale may well be anisotropic with regard to permeability; but water movement depends not only on permeability but also on the existence of a fluid potential gradient. When compaction is due to gravity, the equipotential surfaces in the shale pore water are, ideally, horizontal. The geometry of the shale may disturb this horizontality to some extent, but significant lateral migration requires that the ratio of lateral length of migration path to vertical length be of the same order as the ratio of vertical permeability to lateral permeability. At least for the thinner shales, we may take the vertical path to be half the thickness. So shales of the order of a few hundred metres thick and lateral extent of the order of several kilometres would require a vertical/horizontal permeability ratio of the order of a few thousand, if lateral migration is to be important. While there is great difficulty in estimating the permeability of shales *in situ*, there is

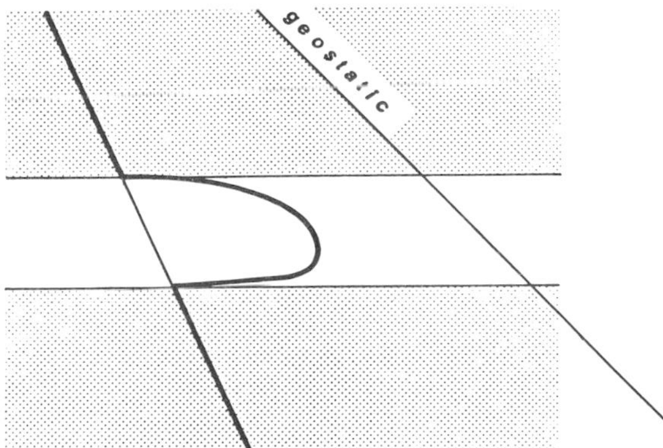

Fig. 6-8. Schematic pressure-depth plot through sand/mudstone/sand sequence.

no evidence to suggest that the transverse to lateral permeability ratio exceeds 10 with any significant frequency – two orders of magnitude too small.

The *quantity* of water expelled depends also on the area normal to the flow. It seems self evident that most shales have a very much larger area normal to vertical migration paths than to horizontal migration paths.

The evidence for vertical migration of shale pore water lies, as we have seen, in the observed and inferred potential gradients that are generated by gravitational compaction. The pore water moves to a space of lower energy. A sand intercalated in a shale may therefore act as a drain, receiving water through its lower surface from below, through its upper surface from above. Within the sand, the water movement must be lateral, probably towards its surface outcrop, or other leakage to the surface. In a typical regressive sequence, in which the sands tend to wedge out in the direction of the regression, the water in the sands will tend to move towards the land-mass from which the regression came. This water can hardly be regarded as connate.

OTHER POSSIBLE CAUSES OF OR CONTRIBUTORS TO ABNORMALLY-HIGH PORE PRESSURES

Many processes have been postulated for the generation of abnormally-high pore pressures – geometry of gas reservoirs, uplift, clay mineral diagenesis, thermal expansion of water, osmosis, and tectonic activity – in addition to the mechanical that we have adopted. There is some merit in each, of course. Let us review them briefly.

Reservoir geometry. There is no doubt whatever that a thick gas column can generate abnormally high pore pressures within the reservoir. This is due to the difference between the normal hydrostatic water pressure gradient outside the reservoir, and the gas pressure gradient within it (Fig. 6-9). This has been recognized from the beginnings of serious study of subsurface pressures. Several reservoirs in Iran have severely abnormal pressures on this account, and a relationship has been observed between reservoir pressures and seeps (see discussion of Dickinson, 1951, pp. 16-17). The effect of thick oil reservoirs is less pronounced, but not always insignificant.

Such abnormally high pressures are particularly difficult to drill successfully because, as can be seen in Figure 6-9, the greatest abnormality is at the top of the reservoir, and the top of the structure, decreasing with depth and with distance from the crest, and vanishing at the gas/water contact.

Uplift. Various authors at various times have suggested that abnormally-pressured shales could result from normally-pressured shales being uplifted to a shallower

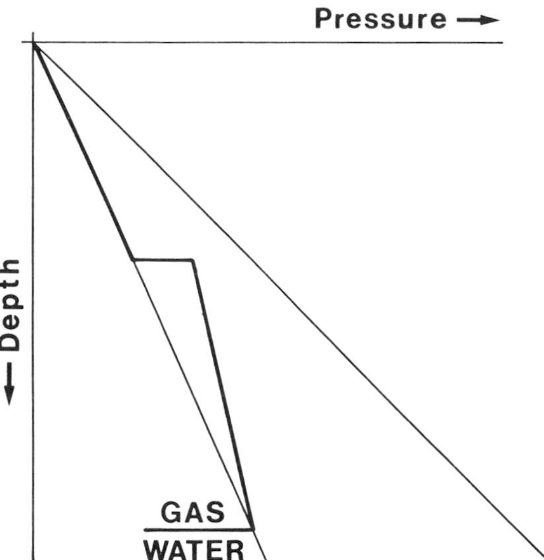

Fig. 6-9. Abnormal pressures developed in gas reservoir by virtue of the density difference between gas and water.

depth without loss of pressure. The main objection to this hypothesis is that there is unambiguous evidence of *subsidence* during the accumulation of regressive sequences in which abnormally-high pore pressures occur*, so this hypothesis must be discarded.

Thermal expansion of water. It has, of course, long been recognized that thermal expansion of water takes place – with some compression – when it is buried ('. . . the expansion of the water at high temperatures exceeds the decrease of volume caused by compression . . .', Versluys, 1932, pp. 924-925), but it could not form part of any hypothesis until the nature of abnormal pressures had been appreciated. This appreciation can be said to have begun in the early 1950's. Thermal expansion of water was suggested as a possible cause of abnormally high pressures by Smith and Thomas (1971, p. 17), argued in more detail by Barker (1972), and then by Margara (1978). It appears to have received wide acceptance, but there are several difficulties.

If thermal expansion is the main cause of abnormally high pressures, it seems that three features are required: the formation of a near-perfect seal to the shale, the maintenance of constant pore volume, and the absence of mechanical compaction. Objections to this hypothesis arise from these apparent requirements.

The most rapid heating of the pore water by burial in the Gulf Coast of the U.S.A. would be about 1 °C per 14,000 years (geothermal gradient of 36 °C/km, burial at

* The accumulation of paralic and neritic sediments in any sequence more than a few hundred metres thick must involve subsidence. The evidence of growth structures will be outlined in the next chapter.

500 years/metre). Thus the maximum rate of thermal expansion of water will be by a factor of 1.0006 in 14,000 years. The intrinsic permeability required to allow the pore water in a column of mudstone 1 m^2 in horizontal cross-section. 500 m deep or thick, and 20% porosity to escape as it expands is of the order of 5×10^{-21} m^2, or 5×10^{-17} cm^2, or about 5×10^{-6} millidarcies (Chapman, 1980). This is near the lower limit of permeabilities measured in Tertiary mudstones in Japan (reported in Magara, 1971, p. 241, fig. 9). Even under the extreme conditions postulated, such permeabilities would be sufficient to dissipate the volume of water generated by thermal expansion. So a near-perfect seal is required.

The maintenace of constant pore volume is intuitively unlikely. As the water is heated by burial, so is the solid material. If this expands isotropically, with maintenance of geometric similarity, the porosity remains unchanged but the pore volume increases. So the pore volume will only remain constant if mechanical compaction reduces the pore volume by the same amount.

That mechanical compaction does take place has already been discussed at some length in the context of Terzaghi's relationship (Chapter 1).

There are other objections. If a perfect seal exists, the pore water pressure gradient will be *hydrostatic*,but displaced to a higher pressure (Fig. 6-10). As with the thick gas reservoirs, the greatest abnormality would be immediately beneath the seal. Drilling and borehole logs indicate the reverse trend: greatest abnormality at the bottom of the transition zone (Fig. 6-11). Also, abnormal pressures occur in places at very shallow depth, above 1,500 m, where the temperature is not great and the prospects of a perfect seal negligible.

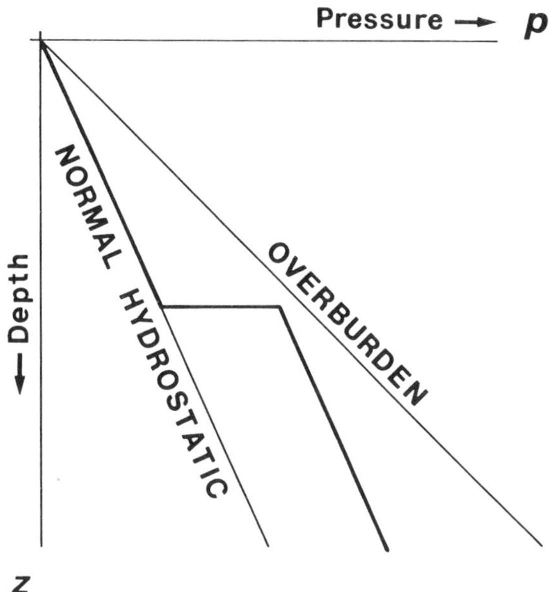

Fig. 6-10. Pressure-depth plot when abnormal pore-water pressures are retained by perfect seal.

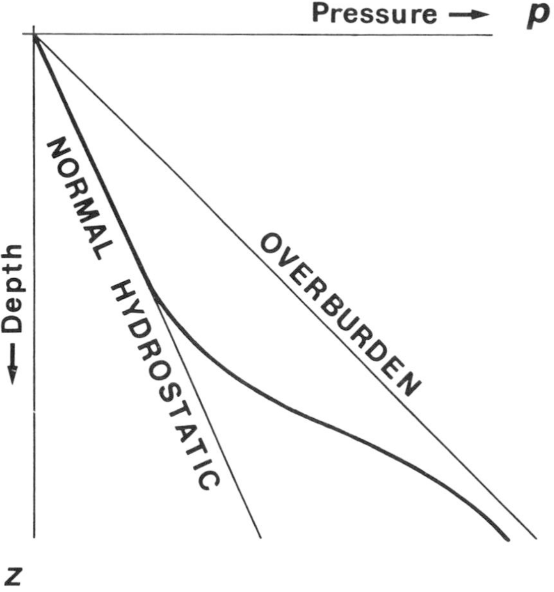

Fig. 6-11. Pressure-depth plot when abnormally-pressured pore water is flowing to positions of lower potential through sediments of low permeability.

Clay mineral diagenesis. This hypothesis stems from a suggestion by Powers (1967) that fluid release from clay minerals could act as a flushing mechanism for petroleum. For this mechanism to generate abnormal pressures, there must be a volume increase when the interlayer water is released to the pore space. Magara (1978, pp. 100-109) has shown that the observed degree of undercompaction could not have been generated by this process, because the expansion, *if* it occurs, is insufficient to account for the bulk density decreases observed.

Furthermore, the diagenesis of montmorillonite (more properly called smectite) to illite, which could release water to the pores, is not regarded by Powers as occurring above a depth of about 1,800 m (6,000 ft), nor by Burst (1969) between 800 and 2,500 m (2,600 to 8,500 ft). Comparable depths were found by Perry and Hower (1970, p. 171) and rather greater depths by Weaver and Beck (1971. p. 18). Abnormal pressures are common at depths shallower than these: for example, Trinidad (Fig. 6-12, which also illustrates normal gradients in permeable beds), New Zealand (Katz, 1974, p. 469), West Irian (Visser and Hermes, 1962, p. 230), Papua New Guinea (Spinks, 1970), Kalimantan, and the Pleistocene of the U.S. Gulf Coast.

While such fluid release may well be a reality at greater depths, the contribution from this cause will only be assessed when we know with confidence whether interlayer water has a higher density than pore water.

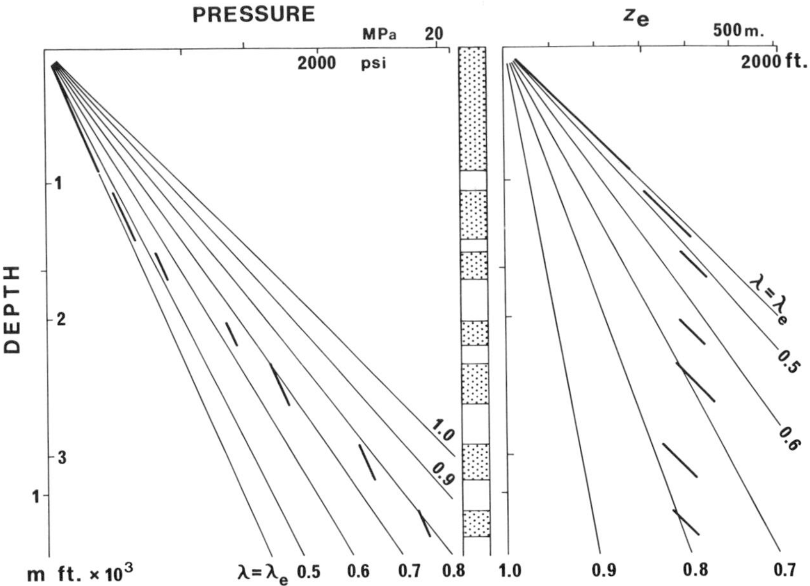

Fig. 6-12. Pressure-depth plot of shallow oil reservoirs in Trinidad, with their corresponding equilibrium depths. (Courtesy of the Royal Dutch/Shell Group.)

Osmosis. We have seen in Chapter 5 that this may be a cause of abnormally low pressures in pore water, and that it could therefore lead to abnormally high pore pressures in the more saline formation to which the water is flowing. It is a matter of contrasting salinities. As a process for the generation of abnormally-high pore pressures in shales, it is unconvincing. Indeed, Margara (1978, pp. 283-284) has argued that the osmotic gradient in a shale opposes the generation of abnormally high pressures by assisting the expulsion of the pore water.

Tectonic. There is ambiguity in those areas of obvious present-day tectonic activity (California, New Zealand) because gravitational influences cannot be eliminated. Nevertheless, when tectonic forces introduce strain in relatively impermeable sediments, this strain is likely to lead to abnormally-high pore pressures rather than a reduction of pore volume, until water can escape. Such a mechanism has been proposed for overthrust faulting (Roberts, 1972) and for the mineralization along faults (Sibson et al., 1975). Tectonic force is probably of little direct significance in many areas of abnormal pressures (such as the U.S. Gulf Coast, the Niger delta, and many areas of S.E. Asia) although, as we have noted, any tendency to slide may induce strain that elevates the pore pressures.

In summary, the mechanical hypothesis may not account for all the abnormality we observe, but it does account qualitatively for the observations. Sediments do tend to compact under load.

There is little doubt that thermal expansion of pore water contributes to abnormal pressures by increasing the volume of water that has to be expelled; but there is considerable doubt that it is significant, let alone dominant.

Clay mineral diagenesis may change the quality of the pore water, but it is doubtful whether it changes the quantity.

There is no doubt at all that thick gas columns can cause seriously abnormal pressures, but this is a different process.

PORE-WATER MIGRATION IN SEDIMENTARY BASINS

We may now combine these various aspects of pore water migration to deduce the general directions of pore water migration in sedimentary basins prior to orogeny. We may take a simple transgressive-regressive cyle as our model (Figure 6-13, which has a grossly exaggerated vertical scale) because the arguments can easily be extended to more complicated sedimentary basins.

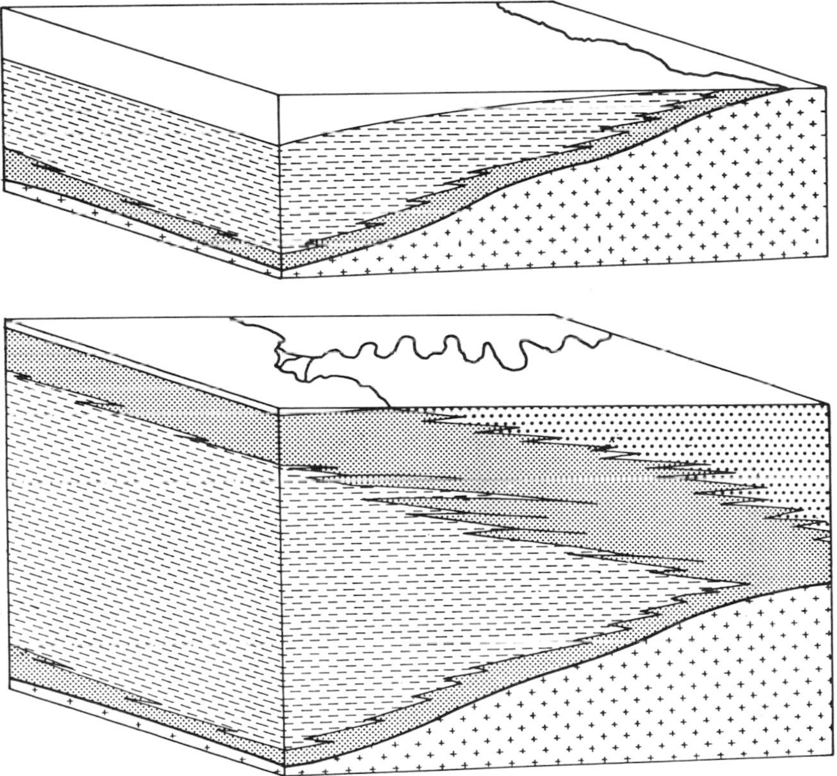

Fig. 6-13. Schematic block diagram of transgressive sequence (upper) and transgressive-regressive cycle (lower).

During the accumulation of the transgressive sequence, the basal permeable unit (commonly a limestone or a sandstone) is in physical continuity with the sea close to the shoreline, and so provides a conduit for the shale pore water that is expelled downwards into it. Much of the compaction of this shale at this stage, though, is seen as an extension of the settling process, with the bulk of the pore water moving upwards relative to the grains. Whether there is absolute upward motion depends on the rate of subsidence relative to the rate of pore-water movement. During the transgressive phase, the shale (or more accurately, mudstone) is compacting from the bottom towards the top.

With the onset of regression, significant changes take place in the pattern of mudstone compaction. Loading the mudstone with a permeable rock unit – in the regressive part of the sequence, almost invariably sand – has the effect of a filter press, and the mudstone beneath it compacts from the top towards the bottom on account of the potential and pressure gradients induced in the upper part of the mudstone. The loss of permeability at the bottom and at the top of the mudstone due to these opposing compaction patterns tends to seal the pore water in the mudstone. Since this is a period of continued subsidence, and therefore of loading by new sediment, the main period of generation of undercompacted, abnormally-pressured 'shales' begins with the onset of regression. This conclusion is supported by the geological evidence of deformation in the regressive sequence during sediment accumulation, due to the mechanical instability of the sequence (this will be discussed in some detail in the following chapter).

The dominant transgressive and regressive sequences characteristically have second-order fluctuations, giving rise to alternating carbonates and shales in the transgressive sequence (typically), or sandstones and shales, and almost exclusively sandstones and shales in the regressive part of the sequence. This alternation is particularly evident in the regressive part of the sequence until paralic and terrestrial sands begin to accumulate. These second-order fluctuations give rise to wedges and tongues of permeable sediment that are discontinuous seaward. The regressive sequence also characteristically exhibits an increasing sand-shale ratio upwards until paralic and terrestrial sands and conglomerates predominate. Harkins and Baugher (1969) noted that in the U.S. Gulf Coast abnormal pressures are typically below the 5% to 10% sand level. These sands, and those above, typically have pore water at normal hydrostatic pressures, from which we deduce that there is hydraulic continuity of pore space to the surface. We infer that pore water is expelled into these sands through both top and bottom interfaces, and that the lateral migration within the sands is, from their geometry, towards the land.

These patterns of pore water migration are of particular importance for the accumulation of petroleum – and, one suspects, of base metals in the transgressive carbonate sequence. Pore-water migration continues throughout the phase of accumulation of sediment in the sedimentary basin, and probably continues well into the orogenic phase. If abnormal pressures exist when the orogenic phase begins, the pore water plays an important role in the nature of the deformation of the sedimentary basin.

SELECTED BIBLIOGRAPHY

Athy, L.F., 1930. Density, porosity, and compaction of sedimentary rocks. *Bull. American Ass. Peroleum Geologists*, 14: 1-24.

Barker, C., 1972. Aquathermal pressuring – role of temperature in development of abnormal-pressure zones. *Bull American Ass. Petroleum Geologists*, 56 (10): 2068-2071.

Burst, J.F., 1969. Diagenesis of Gulf Coast clayey sediments and its possible relation to petroleum migration. *Bull. American Ass. Petroleum Geologists*, 53: 73-93.

Chapman, R.E., 1972. Clays with abnormal interstitial fluid pressures. *Bull American Ass. Petroleum Geologists*, 56 (4): 790-795.

Chapman, R.E., 1973. *Petroleum geology: a concise study*. Elsevier Scientific Publ. Co., Amsterdam, London, and New York, 304 pp.

Chapman, R.E., 1976. *Petroleum geology: a concise study* (paperback edition). Elsevier Scientific Publ. Co., Amsterdam, Oxford, and New York, 302 pp.

Chapman, R.E., 1979. Mechanics of unlubricated sliding. *Bull. Geol. Soc. America*, 90: 19-28.

Chapman, R.E., 1980. Mechanical versus thermal causes of abnormally high pore pressure in shales. *Bull. American Ass. Petroleum Geologists*, 64: 2179-2183.

Dailly, G.C., 1976. A possible mechanism relating progradation, growth faulting, clay diapirism and overthrusting in a regressive sequence of sediments. *Bull. Canadian Petroleum Geology*, 24 (1): 92-116.

Dickinson, G., 1951. Geological aspects of abnormal reservoir pressures in the Gulf Coast region of Louisiana, U.S.A. *Proc. 3rd World Petroleum Congress* (The Hague, 1951), sect. 1: 1-16. (Discussion: 16-17.)

Dickinson, G., 1953. Geological aspects of abnormal reservoir pressures in Gulf Coast Louisiana. *Bull. American Ass. Petroleum Geologists*, 37 (2): 410-432.

Fertl, W.H., 1976. *Abnormal formation pressures*. Elsevier Scientific Publ. Co., Amsterdam and New York, 382 pp. (Developments in Petroleum Science, 2).

Füchtbauer, H., and Reineck, H.-E., 1963. Porosität und Verdichtung rezenter, mariner Sedimente. *Sedimentology*, 2: 294-306.

Harkins, K.L., and Baugher, J.W., 1969. Geological significance of abnormal formation pressures. *J. Petroleum Technology*, 21: 961-966.

Hedberg, H.D., 1926. The effect of gravitational compaction on the structure of sedimentary rocks. *Bull. American Ass. Petroleum Geologists*, 10 (11): 1035-1072.

Hedberg, H.D., 1936. Gravitational compaction of clays and shales. *American J. Science*, 5th series, 31: 241-287.

Hottman, C.E., and Johnson, R.K., 1965. Estimation of formation pressures from log-derived shale properties. *J. Petroleum Technology*, 17: 717-722.

Hubbert, M.K., and Rubey, W.W., 1959. Role of fluid pressure in mechanics of overthrust faulting, I. Mechanics of fluid-filled porous solids and its application to overthrust faulting. *Bull. Geol. Soc. America*, 70 (2): 115-166.

Katz, H.R., 1974. Recent exploration for oil and gas. *In:* G.J. Williams (Ed.), Economic geology of New Zealand. *Monograph Series Australasian Inst. Mining Metallurgy*, 4: 463-480.

Magara, K., 1971. Permeability considerations in generation of abnormal pressures. *J. Soc. Petroleum Engineers*, 11: 236-242.

Magara, K., 1978. *Compaction and fluid migration: practical petroleum geology*. Elsevier Scientific Publ. Co., Amsterdam, 319 pp. (Developments in Petroleum Science, 9).

Maxwell, J.C., 1964. Influence of depth, temperature, and geologic age on porosity of quartzose sandstone. *Bull American Ass. Petroleum Geologists*, 48 (5): 697-709.

Perry, E., and Hower, J., 1970. Burial diagenesis in Gulf Coast pelitic sediments. *Clays and Clay Minerals*, 18: 165-177.

Plumley, W.J., 1980. Abnormally high fluid pressure: survey of some basic principles. *Bull American Ass. Petroleum Geologists*, 64 (3): 414-422.
Powers, M.C., 1967. Fluid-release mechanisms in compacting marine mudrocks and their importance in oil exploration. *Bull. American Ass. Petroleum Geologists*, 51 (7): 1240-1254.
Roberts, J.L., 1972. The mechanics of overthrust faulting: a critical review. *Proc. 24th International Geological Congress* (Montreal, 1972), Sect. 3: 593-598.
Rubey, W.W., and Hubbert, M.K., 1959. Role of fluid pressure in mechanics of overthrust faulting, II. Overthrust belt in geosynclinal area of western Wyoming in light of fluid-pressure hypothesis. *Bull. Geol. Soc. America*, 70 (2): 167-206.
Sibson, R.H., Moore, J.McM., and Rankin, A.H., 1975. Seismic pumping – a hydrothermal fluid transport mechanism. *J. Geol. Soc. London*, 131: 653-659.
Smith, J.E., 1971. The dynamics of shale compaction and evolution of pore-fluid pressures. *Mathematical Geology*, 3: 239-263.
Smith, J.E., 1973. Shale compaction, *J. Soc. Petroleum Engineers*, 13: 12-22.
Smith, N.E., and Thomas, H.G., 1971. Origins of abnormal pressures. *In:* Houston Geol. Soc., *Abnormal subsurface pressure: a Study Group report* 1969-1971. Houston Geol. Soc., Houston, pp. 4-19.
Spinks, R.B., 1970. Offshore drilling operations in the Gulf of Papua. *J. Australian Petroleum Exploration Ass.*, 10: 108-114.
Stephenson, L.P., 1977. Porosity dependence on temperature: limits on maximum possible effect. *Bull American Ass. Petroleum Geologists*, 61 (3): 407-415.
Terzaghi, K., 1936. Simple tests determine hydrostatic uplift. *Engineering News Record*, 116 (June 18): 872-875.
Versluys, J., 1932. Factors involved in segregation of oil and gas from subterranean water. *Bull American Ass. Petroleum Geologists*, 16 (9): 924-942.
Visser, W.A., and Hermes, J.J., 1962. Geological results of the exploration for oil in Netherlands New Guinea. *Verhandelingen van het Konnklijk Nederlandsch Geologisch-Mijnbouwkundig Genootschap voor Nederland en Kolonieën*, Geologische Serie, 20: 1-265.
Weaver, C.E., and Beck, K.C., 1971. Clay water diagenesis during burial: how mud becomes gneiss. *Special Paper Geol. Soc. America* 134 (96 pp.).
Weller, J.M., 1959. Compaction of sediments. *Bull. American Ass. Petroleum Geologists*, 43: 273-310.

7. ROLE OF PORE WATER IN DEFORMATION OF SEDIMENTARY BASINS

We must make the distinction between pre-orogenic deformation and orogenic deformation of sedimentary basins because there is a great deal of evidence that important deformation occurs in sedimentary basins while they are accumulating sediment on a subsiding sedimentary column – particularly in the terminal regressive sequence. The regressive sequence itself is due to a neighbouring orogeny outside the sedimentary basin: the creation of mountains creates sediment in increasing amounts until the supply of sediment to a sedimentary basin exceeds the volume created by subsidence of the sedimentary basin. The sea then tends to become shallower, and the sediments prograde away from the orogeny.

Deformation in this part of the sedimentary basin that takes place while sediment is accumulating is here referred to as *pre-orogenic* because it takes place before orogeny of the sedimentary basin itself. (The term is not exclusive to this situation: other parts, particularly the initial transgressive sequence also suffer pre-orogenic deformation – but we are concerned with the role of pore water.) We shall later discuss orogenic deformation, which takes place during orogeny of the sedimentary basin itself. The relationship between the two is shown schematically in Figure 7-1. As a modern example, we may take the U.S. Gulf Coast basin. Throughout the Cenozoic, it has been accumulating sediment derived from the relatively distant Rocky Mountains and Appalachians. It has not yet suffered orogeny, but it has suffered extensive faulting and gentle folding apart from that due to salt diapirism.

We shall first consider the geology of pre-orogenic deformation, then its probable causes.

PRE-OROGENIC DEFORMATION

Growth structures

Structures that formed while the sediment in them accumulated are known collectively as *growth structures*. *Growth faults* are faults (usually normal faults in the context of regressive sequences, but not uniquely so) across which the thicknesses of correlative rock units change abruptly (Fig. 7-2), the thicker being on the downthrown side. Likewise, *growth anticlines* are anticlines in which the rock units are

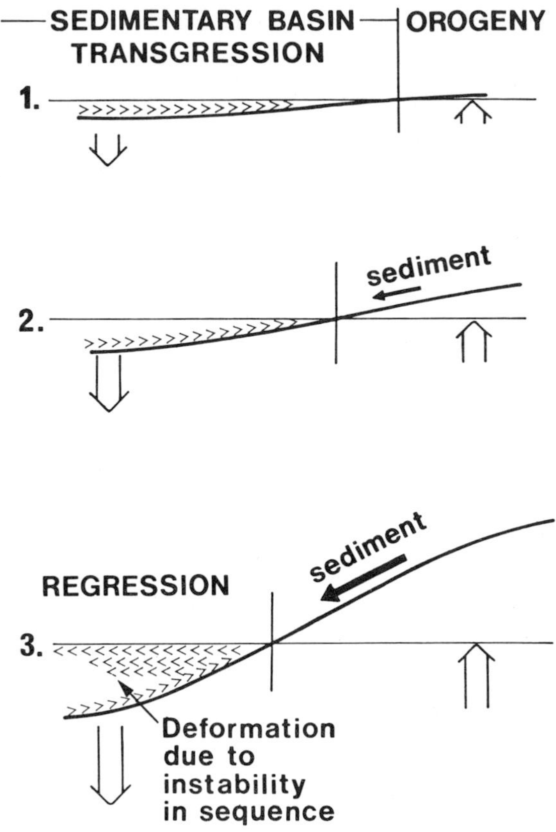

Fig. 7-1. Schematic relationship between orogenic and pre-orogenic deformation.

thinner over the crest than on the flanks (Fig. 7-3). In general, growth structures are structures in which the thicknesses of rock units reflect the structure itself. Sedimentary basins are growth structures on a large scale.

The first recognition of growth faults appears to be due to Tiddeman (1890), who ascribed the lithological changes across the Craven fault in England to movement of the fault *concurrently* with sediment accumulation. Subsequently, growth structures were found almost exclusively in the domain of applied geology, first in coal mines, then in oil and gas fields. By their nature, they are more easily recognized in subsurface workings, with detailed data in three dimensions, than in outcrop. Dron (1910) found such faults in the Scottish coal fields, and the existence and interpretation of complicated growth structures became well accepted in the coal fields of western Germany in the 1920s (Böttcher, 1925, 1927), later to be extended to European oil fields (Stutzer, 1930).

They were rediscovered a decade or so later by the U.S. petroleum industry, and given various names: *Gulf Coast Type fault* (!), *synsedimentary*, *progressive*, *deposit-*

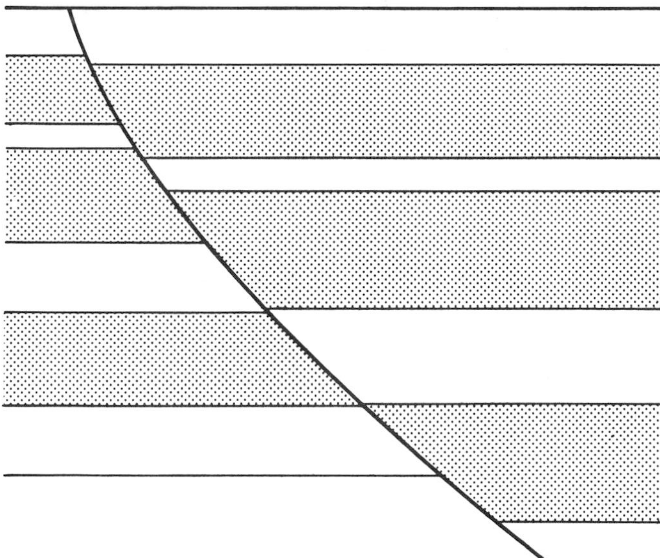

Fig. 7-2. Cross-section through typical growth fault.

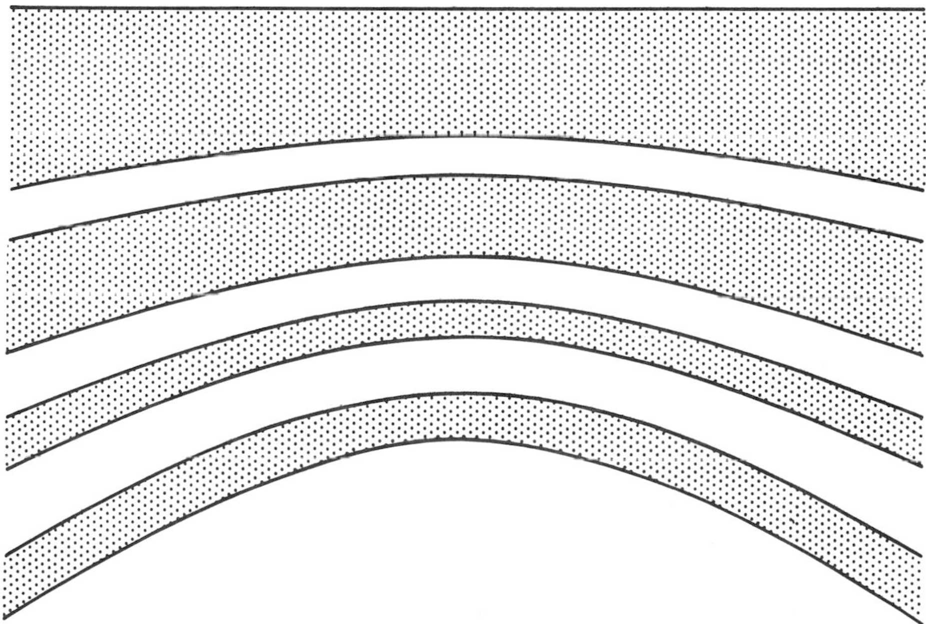

Fig. 7-3. Cross-section through typical growth anticline.

ional, *contemporaneous* and *growth*. The terms *depositional* and *growth* were in common use in the petroleum industry in the 1950s, but it was *growth* that was recommended by Dennis (1967) in the *International Tectonic Dictionary*. It would be hard to make a case for priority for *growth*, but it has its merits and we shall use it throughout.

The interpretation of growth structures as structures that grew while the sediment accumulated was clearly reached quite independently by the German coal geologists and the U.S. petroleum geologists (there is no evidence that the U.S. writers had read the German works – or, indeed, that either had read Tiddeman's paper).

Growth faults occur in two main stratigraphic associations: the initial transgressive sequence of sedimentary basins formed by rifting and basement faulting (detected in recent years by marine seismic surveys), and the terminal regressive sequence of sedimentary basins. In the latter, we have the coal-field occurrences already noted (the major coal-bearing sequences of the world are dominantly regressive), and the oil and gas field occurrences, which are also associated with shales with abnormally high pore-water pressures. We are concerned mainly with the latter.

Growth faults in regressive sequences are typically slightly curved in plan (concave to the direction of progradation of the regressive sequence) and curved in section (concave up). The strike of the fault is roughly parallel to the depositional strike, with the downthrown side on the side that is in the direction of progradation*. The dip of the fault typically flattens with depth from about 60° or 70° to 40°, or even less.

It seems to the writer that considerable misunderstanding has developed over this flattening. Because abnormal pressures are found in the U.S. Gulf Coast in the flatter part of a growth fault, and these pressures increase with depth as the fault flattens, it has been implied (e.g., Bruce, 1973, p. 884) that the two observations have a causal relationship. Apart from the theoretical and practical evidence that pore pressures do not affect the *angle* of fracture (see Hubbert and Rubey, 1959, pp. 141-142), it must be remembered that the angle of fault dip was determined at the time the fault was initiated, which was soon after the beginning of accumulation of the permeable sands of the regressive sequence – that is, in unconsolidated sediments at shallow depth. This angle is typically about 60° to the bedding, but compaction of the sediments (by definition, after the fault has been created) tends to flatten the dip. Since the proportion of shale increases downwards in a regressive sequence, so the relative compaction increases and flattens the dip of the fault. If the original angle was 60° and the original porosity 50%, 40% compaction of a faulted shale to 16% porosity will reduce the dip of the fault to 45°. If the original fault dip was 50°, 40% compaction will reduce it to 35°.

So most of the flattening can be attributed to compaction. It is most unlikely that

* This is the characteristic that leads to the objectionable name 'down-to-the-basin fault'. It is objectionable because it implies a position on the edge of a 'basin' and reveals a misunderstanding of the nature of sedimentary basins.

the fracture can be sustained in the undercompacted shale; but flow of shale away from the loaded area may account for some extreme flattening reported. The rate of movement of a growth fault is of the order of 100 m per million years or so – 10^{-4} m per year – which is an infinitessimal movement for plastic shale.

One must also remember that any tilting or folding of the beds tilts growth faults equally (see Murray, 1961, p. 183, fig. 4.27, for an example).

The movement of growth faults can be dated relative to the sediments they cut, because thickness contrasts are only generated while the fault is moving. Plotting the thickness ratio of correlative rock units across the fault against the stratigraphic level dates the movement of the fault and its relative activity. No worldwide study of the time of faulting relative to the transition zone to abnormal pressures has been published, but Thorsen (1963) made an interesting study of growth faults in western Louisiana.

Comparison of Thorsen's map (1963. p. 107, fig. 4) with Dickinson's (1953, pp. 416-417, fig. 3) shows that the periods of *maximum* growth-fault movement occurred in Louisiana within a couple of biostratigraphic zones above the youngest abnormally-pressured shales. Thorsen also noted that the time of maximum growth-anticline movement generally was the same as for the faulting. This indicates that growth-structure movement began soon after the accumulation of the more permeable part of the sequence on top of the shales, and suggests that abnormal pressures also developed early in these shales when the overburden was relatively slight (cf. equilibrium depth in previous chapter). This is also supported by Thorsen's observation that sand percentage is most closely related to contemporaneous structural growth near 'the basinward limit of sand deposition, that is, in those areas of ten per cent or less sand', and Harkins and Baugher's (1969) observation that the top of abnormally-pressured shale normally occurs regionally below the 5-10% sand level.

In the Midland field, studied by Fowler et al. (1971, p. 65), faulting began soon after the permeable part of the sequence began to accumulate. The writer has seen evidence of fault movement with less than 500 m of overburden on what are now abnormally-pressured shales.

There is thus strong field evidence to support the theoretical arguments for early generation of abnormally-pressured shales in regressive sequences, and the early generation of all those parameters related to such abnormal pressures.

When more than one growth fault is found in an area, they tend to be younger in the direction of progradation, so that growth faults on the downthrown side of a growth fault tend to be younger that those on the upthrown side. They tend to begin their movement and to stop their movement later in the direction of progradation, so that the stratigraphic relationships of a sequence of growth faults may be as in Figure 7-4. For examples of this, see Murray (1961, p. 190, fig. 4.33) and compare with Dickinson (1953, pp. 416-417, fig. 3, and p. 419, fig. 5). Figure 7-5 illustrates this relationship between growth faults and other features.

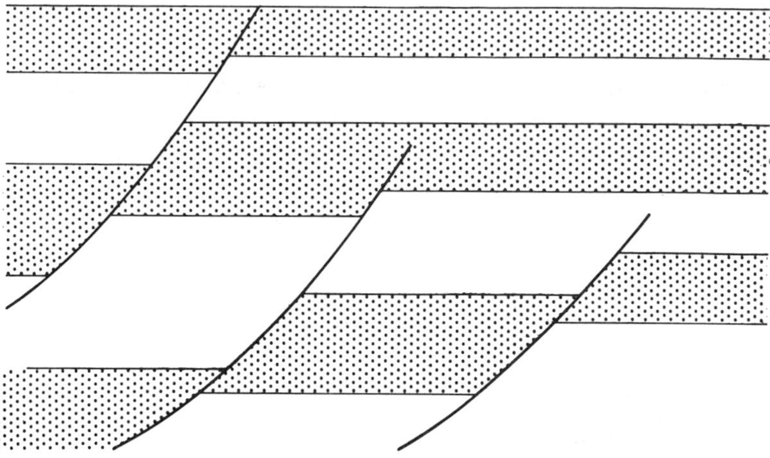

Fig. 7-4. Cross-section through typical sequence of growth faults.

Growth anticlines, in which the thicknesses of rock units decrease towards the crest, are mapped by the construction of isopach and cumulative isopach maps (see Chapman, 1973 or 1976, pp. 221-223). The cumulative isopach contours for deeper horizons (larger intervals) have a configuration resembling that of the structural contours of that horizon. As a consequence of these thickness changes, growth anticlines tend to become steeper-flanked with depth. This feature appears to have been noted first by Stutzer (1930). Relatively little attention has been given to them in the literature and, apart from Thorsen (1936) no explicit data has been published on the timing of initial growth in relation to the overburden on the main shale part of the sequence. The writer's experience agrees with Thorsen's, that the main period of anticlinal growth embraces that of growth-fault movement, but that anticlinal growth may continue after growth-fault movement has ceased. In other words, an individual growth fault may have a shorter history of movement than its associated anticline.

Causes of growth structures

The search for the causes of growth structures must begin with a mechanical analysis. The mechanics of faulting is well understood, and there is no ambiguity about the evidence that a normal fault develops in a stress field in which the least and intermediate compressive stresses are in the horizontal plane, and the greatest compressive stress is vertical (we take these principal stresses to be horizontal and vertical because it is a close approximation under the continental shelves and low lying coastal areas, where these phenomena are commonly found). A normal fault makes an angle of about 30° with the axis of greatest principal stress, that is, its dip is about 60°; and the strike of the fault is normal to the axis of least principal stress.

We speak of 'least' and 'greatest' compressive stresses because in unconsolidated sediment with no tensile strength, all sediment below a few metres from the surface is in compression (Hubbert, 1951, p. 367). But there is a *component* of tension (the deviatoric stress of structural geologists) normal to the strike of a normal fault, in the direction of lateral displacement: this component of tension is usually aligned roughly normal to the depositional strike.

Thus growth faulting is seen, at least in part, as resulting from a tendency for the materials to slide seawards in such open-ended situations as the U.S. Gulf Coast and the Niger delta. We shall see later that bulk movement of the shale outward from under the load may also contribute to growth faulting.

When growth faults are associated with growth anticlines, we may extend the argument significantly. If growth anticlines were growing while the growth faults in them were moving, then the growth anticline formed in the same stress field as the growth faults, that is, *with the greatest principal stress vertical*. This conclusion is at variance with orthodoxy, which would require the axis of greatest compressive stress to be horizontal and normal to the axis of the fold. So it must not be accepted without further scrutiny*. We therefore take a field example of which the data are not in dispute – Seria oil field in Brunei.

The Seria field is an asymmetric anticline (Fig. 7-5) in Tertiary sediments. It is about 20 km long and less than 5 km wide, with the axis parallel and close to the present-day coastline (Schaub and Jackson, 1958; Liechti et al., 1960). The stratigraphic sequence is regressive, passing upwards from the neritic Setap Shale Formation (Oligo-Miocene), through the neritic shale/sand Miri Formation and the neritic sand/shale Seria Formation, to the paralic Liang Formation. The regression is apparently still going on, to the north or north-west.

The structure has no surface expression: it was revealed by shallow core-drilling under a flat, low-lying, coastal swamp. The drilling of several hundred wells has revealed beyond reasonable doubt that it is a growth anticline cut by growth faults, some of which cut the youngest sediments. So the evidence is that the anticline was formed in a stress field with a horizontal component of tension, the greatest principal stress being vertical.

Schaub and Jackson (1958) attributed this folding of the Northwest Borneo basin to orogeny towards the end of the Pliocene. They noted that while the folding is more intense in the interior of Borneo, it decreases 'basinward' where broad gentle synclines are separated by narrow, steep anticlines (of which Seria is one) that are complexly faulted by normal faults in the main. These faults developed contemporaneously with the accumulation of the younger Neogene.

It will be noted that if the folding that is ascribed to a Pliocene orogeny implies horizontally-directed compressional deformation, then the field data do not support such an hypothesis because normal faulting was taking place in the same

* They are not caused by the folding because they are not uniquely associated with the folding; but folding may, of course, contribute to the faulting.

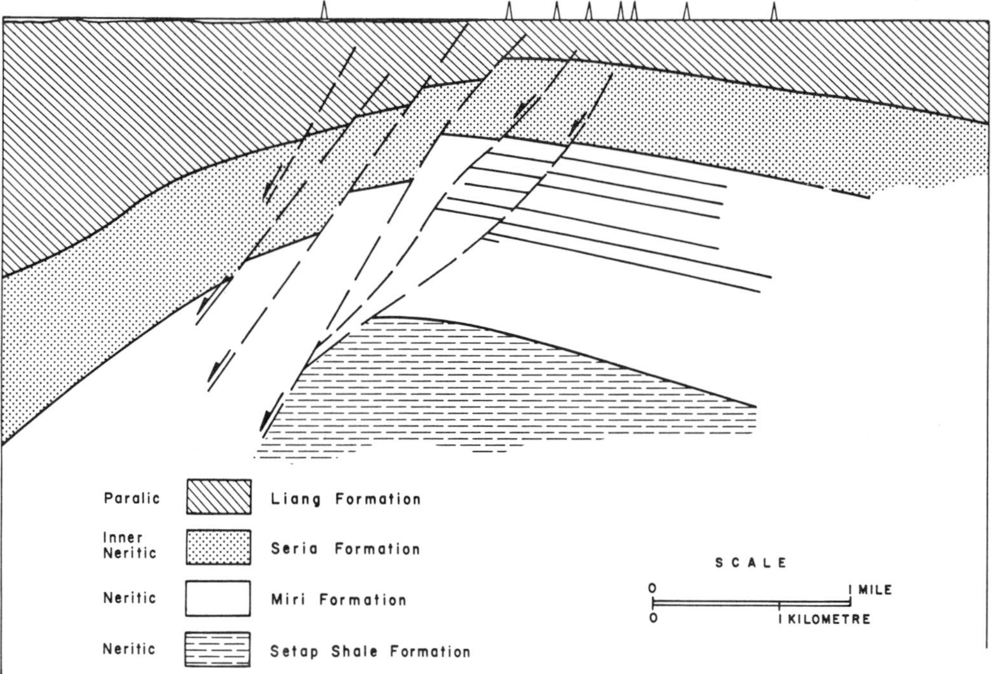

Fig. 7-5. Cross-section through Seria field, State of Brunei (after Schaub and Jackson, 1958, fig. 3).

sediments in the same place at the same time. Lest it be thought that the reverse fault in Miri field (Schaub and Jackson, 1958, p. 1332, fig. 2) supports the hypothesis of horizontal compressive deformation, it must be pointed out that a reverse fault dipping at about 60° is consistent with *vertical* deforming forces because fractures make an acute angle with the axis of greatest principal compressive stress.

The following observations appear to be significant: a) The youngest formation, the Liang, is folded (it was discerned from the original shallow core drilling). b) This formation is also faulted by normal faults, and c) even over the crest, there are more than 2,000 m of neritic and paralic sediments.

These observations surely lead us to the conclusion that the deformation of Seria field took place during the Miocene to Holocene (at least) during *subsidence* in a stress field with a component of *horizontal tension*, and the orthodox view cannot be sustained.

Let us assemble all the pertinent data, to help us to a better understanding of Seria (and similar structures):

- The stratigraphic sequence is clearly dominantly regressive.
- The folding and faulting took place while the sediment accumulated and subsided.
- The anticline is very gentle at shallow depth, but increases in steepness with depth. Hence deformation continued to the very recent past, at least.

- The Setap Shale is abnormally pressured in the Seria and Miri areas, and this is probably a regional feature because it forms mudvolcanoes in the Jerudong-Limbang area of Brunei and at Bulak Setap (from which it takes its name) (Liechti et al., 1960, pp. 326-329).
- The structural style is one of broad, gentle, synclines separated by narrow, steep, anticlines.

These features – regressive sequence, growth structures, abnormally-pressured shale – are found in many parts of the world in conjunction with this structural style. It has been argued extensively that this association has important relations with mechanical instability in young regressive sequences (Chapman, 1973, 1976, 1977a, 1977b). It is a diapiric* deformation.

The theory of deformation due to mechanical instability in simple sequences is quite well known for simple geometry (Biot and Odé, 1965; Ramberg, 1972, for example). Scale models agree well with theory (Parker and McDowell, 1955; Ramberg, 1967; Tanner and Williams, 1968); and both the theoretical and physical models resemble geological observations closely. A two-layered sequence in which the upper layer is denser than the lower is mechanically unstable in the field of gravity, and the interface will deform if the viscosities are sufficiently low. In particular, Tanner and Williams (1968) found that anticlines tended to form with their axes parallel to the intermediate principal stress.

When we speak of *viscosity* in rocks, we should qualify it with an adjective such as *equivalent* or *apparent* to indicate that we are considering a property analogous to viscosity in liquids. We must assume that in geology the strict analogy is with several immiscible liquids of different viscosities – and therein lies the difficulty of quantifying this property of rocks. In the present context the analogy is broadly justified on the grounds that properly scaled physical models use materials that resemble liquids more closely than the materials they represent, because of the dimension of time. Some workers have used liquids and obtained results like those of nature (e.g., Ramberg, 1967, p. 53ff, using oils). Qualitatively, we must accept the field evidence that materials that are brittle on the short time-scale have been folded without fracture on the long time-scale (we see this property also in pitch, as in the Trinidad Pitch Lakes, and sealing wax).

If we regard a young regressive sequence as a two-layer system of permeable sands with normally-compacted shales overlying undercompacted over-pressured shales, we are concerned with the relative viscosities of these two layers. The experience of drilling for petroleum, as we saw in the previous chapter, indicates that the physical properties of these two layers are very different, and that the pore-water pressure is the significant parameter leading to these differences. When drilling through the upper, normally-pressured layer, sands typically drill faster than shales, but the hole

* Purists are quite right to insist that a diapir is penetrative: but it is surely permissible to use 'diapiric' for analogous processes. Salt diapirs are preceded by a non-penetrative deformation into salt waves. This is the analogy.

remains stable through both. As soon as the transition zone is reached, the driller observes what is known as a 'drilling break' – the rate of penetration increases rapidly and soon exceeds that experienced in sands although now drilling in shales. The two main reasons for this, both related to pore pressure, are the reversal or reduction of the fluid potential gradient across the bottom of the hole, and the reduction of the competence or coherence of the shale due to the reduction of effective stress as the pore pressures increase above normal hydrostatic.

When undercompacted shales are encountered at relatively shallow depth (above 2,000 m approximately) they commonly tend to flow into the borehole ('heaving shales', or 'gumbo' in the U.S. Gulf Coast). This effect is detected by increased torque while drilling, or failure to reach bottom when going into the hole after changing the bit. Electrical logs with a caliper for measuring the diameter of the hole (such as the Microlog) may reveal shales sections with a hole diameter less than the diameter of the bit that drilled the hole.

There is thus little doubt that abnormally-pressured shales that are undercompacted also have a lower viscosity than normal for their depth. This conclusion is supported by the evidence of mudvolcanoes, which occur in the same association with young regressive sequences, and are true diapiric phenomena.

The matter of bulk density of these sediments is elusive. As we have seen, the bulk wet density of a sedimentary rock can be expressed in terms of the mass densities of the constituents and the porosity (all of which can be measured or closely estimated),

$$\begin{aligned}\rho_{bw} &= f\rho_w + (1-f)\rho_s, \\ &= \rho_s - f(\rho_s - \rho_w), \end{aligned} \tag{7.1}$$

where f is the porosity, ρ_w the mass density of the pore water, ρ_s that of the solids. It follows, therefore, that where the porosity is increasing with depth in the transition zone to an undercompacted shale, so the bulk wet density of the shale decreases. If a functional relationship can be found between depth, porosity, and pore-water pressure, then the expression above can be modified to take these parameters into account. Athy's formula (equation (6.2)) can be generalized by the introduction of equilibrium depth (the parameter δ), and using this as the most unfavourable curve for comparative purposes, we may substitute

$$f = f_0 e^{-c\delta z} \tag{7.2}$$

into equation (7.1), obtaining

$$\rho_{bw} = \rho_s - f_0 e^{-c\delta z} (\rho_s - \rho_w). \tag{7.3}$$

This relationship is plotted in Figure 7-6 for $\rho_s = 2{,}650$ kg m^{-3}, $f_0 = 0.48$, $c = 1.42 \times 10^{-3} m^{-1}$, $\rho_w = 1{,}050$ $kg\ m^{-3}$; and from it we see that if the mean overburden bulk wet density is 2,300 kg m^{-3}, density inversions can exist in abnormally-pressured shales down to depths of at least 2,000 m.

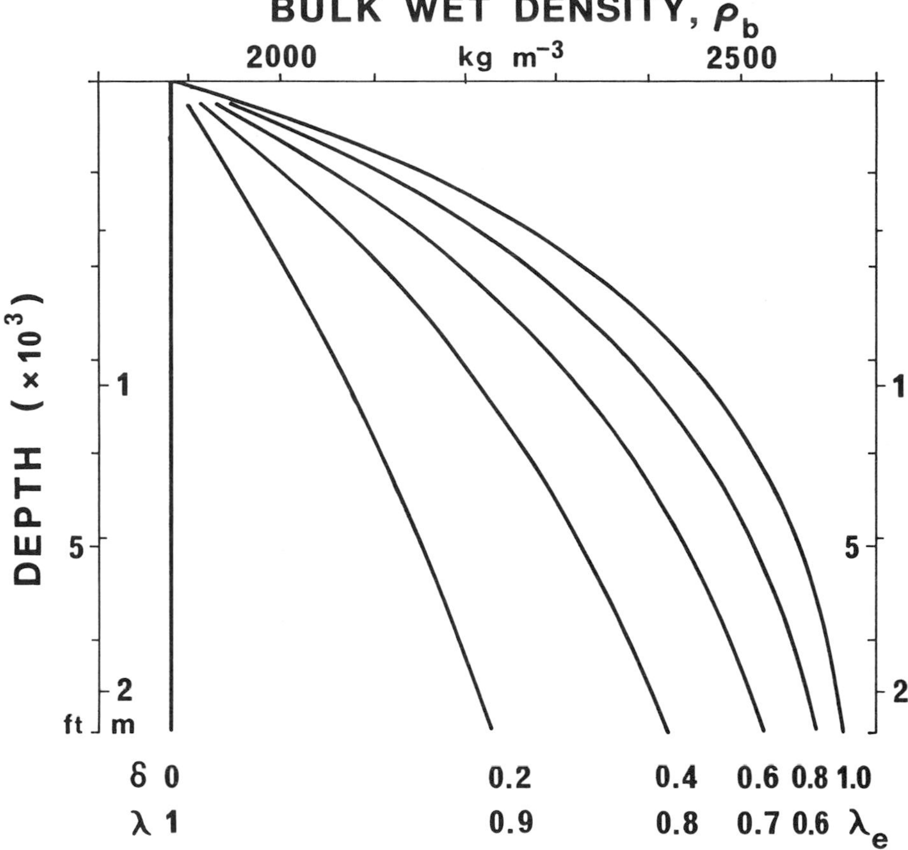

Fig. 7-6. Idealized relationship between depth, pore-fluid pressure (through the parameters δ and λ) and bulk wet density of a sediment.

From a practical point of view, there are several borehole logs that yield a measure of bulk density of the rocks in the wall of the borehole (one problem with these is that the correction factor to be applied is often a significant proportion of the number obtained). Nevertheless, density inversions are revealed by such logs when run in boreholes with abnormal pore pressures in the shales, and they can be used for estimating the values of f_0 and c in equation (7.3). (There is a linear relationship between bulk density and porosity; plot the natural logarithm of porosity against depth and the slope gives the value of c, and the value of f_0 is obtained by extrapolating the straight-line part to $z = 0$.)

The rate of deformation of an unstable sequence, and the geometry of the deformation, depend on the density contrast between the two layers, their viscosity contrast, and the thicknesses. It is found that the earliest deformation of the interface between the two layers is of short wavelength and low amplitude, but, as the deformation proceeds, so one wavelength is amplified at the expense of the others. This is known as the *dominant wavelength* (see Biot and Odé, 1965). It is a function of the viscosity ratio and the thickness ratio of the upper and lower layers. The higher the viscosity of the overburden relative to that of the lower layer, the longer the dominant wavelength and the slower the rate of deformation. Biot and Odé found that with overburden/mother-layer thickness ratios between 1 and 5, there was a significant rate of growth with a dominant wavelength about 10 to 20 times the thickness of the mother layer.

The main physical ingredients of gravitational instability – density inversion and relatively low viscosity – are present in young regressive sequences with abnormally-pressured shales. Also, the structural style so commonly observed – broad gentle synclines separated by narrow steep anticlines – is in harmony with the development of a dominant wavelength, as revealed by mathematical and physical models with unstable sequences. Thus the argument for vertical rather than horizontal deforming forces is supported by a rigorous qualitative argument around the field data of growth structures; and the key to this argument is the role of pore water in influencing the physical properties of the sediments. The problems of quantifying the argument have not yet been solved. These difficulties relate largely to the dimension of time, and the manner in which the physical dimensions and properties of a three-dimensional geological space change with time.

With the passage of time, subsidence and the accumulation of sediment change the geometry of the rock units seen as physical units; and the expulsion of pore water changes both the densities and the viscosities of the rocks with time. These variables are not amenable as yet to analysis. But it is worthwhile noting that these processes have a general tendency to reduce the degree of mechanical instability with time, and sooner or later the sequence will become stable. When or if the main shale wedge has reached compaction equilibrium, it will be more dense than the overburden. These tendencies may account for the observation that salt domes are much more common than true shale diapirs, because the dimension of time has little or no effect on the physical properties of salt.

Finally, it will occur to the reader that certain 'open-ended' stratigraphic sequences will suffer a deformation due to the load on a relatively plastic mass of shale. Such a situation, depicted in Figure 7-7, is relevant to the U.S. Gulf Coast and many of the major deltas. The load will tend to squeeze the shale out laterally, generating a bulk flow.

Dailly (1976) has presented a cogent argument for this occurrence in the Niger delta; and it is clear that such outward flow may well contribute to the development of normal growth faults, and that sliding in not the only possibility. On a small scale, processes such as these have been studied in the Mississippi delta, where the generation of 'mudlumps' has been attributed to an overburden load of less than 100 m thickness. Folding and overthrusting were found in core-holes (Morgan et al., 1968; but see also Lyell, 1867, pp. 447-454 for what is probably the earliest interpretation of mudlumps).

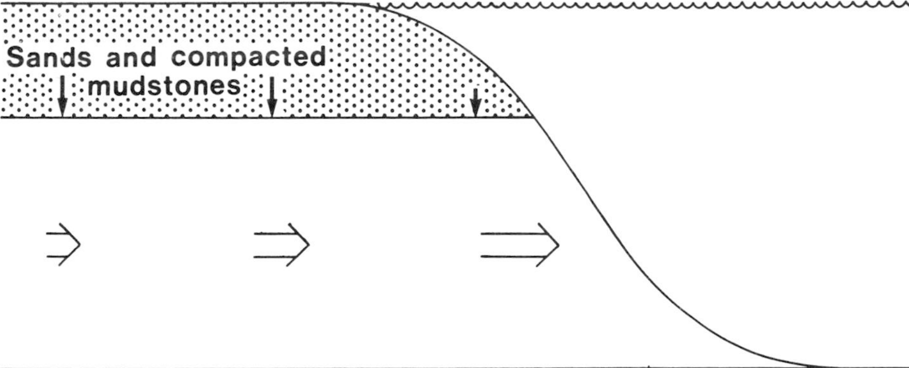

Fig. 7-7. Schematic bulk movement of abnormally-pressured shale as a result of overburden load in a delta.

OROGENIC DEFORMATION*

If orogeny of the sedimentary basin itself begins before compaction equilibrium has been achieved, further deformation may take place in which the pore water plays an important role. It is evident that this will be superimposed on any pre-orogenic deformation, and that this superimposition may be a decisive factor in the style of orogenic deformation. Most obviously, the creation of slopes may lead to sliding (and equally obviously, there may be no hard and fast line to be drawn in time between slopes created by basin subsidence and those created by the onset of orogeny).

The previous argument included the matter of relatively low viscosity in the abnormally-pressured shales: such shales will act as a lubricant for the sliding of a regressive sequence, or any sequence that has the same physical characteristics.

Lubricated sliding

When a block is placed on a layer of material of uniform lower viscosity, and the whole tilted, deformation of the less viscous underlying layer results in movement of the block down the slope. For the same block on the same slope, the lower the viscosity of the lubricant, the more rapid the movement. For the same block on the same lubricant on the same slope, the thicker the lubricant, the faster the movement. The steeper the slope, of course, other things being the same, the more rapid the movement.

Kehle (1970) examine lubricated geological sliding in some detail in an interesting paper, and concluded that such a mechanism could easily account for most cases of

* We use the term 'orogeny' in its original sense of 'mountain-building', not the modern variant connected with the emplacement of granitic plutons.

tectonic translation. In spite of (or perhaps, because of) the simplifying assumptions, the computed velocities of sliding were far in excess of those required to account for observed translations. We shall look first at some field examples of deformation that have been attributed to sliding.

In western Papua New Guinea (Fig. 7-8), the Mesozoic and Tertiary sequence of the Southern Highlands basin, south of the Kubor Range, appears to have slid southwards as the Kubor Range was elevated, with deformation taking the form of thrusting and overthrusting, and steep, asymmetric folding (Jenkins, 1974; Findlay, 1974; Ridd, 1976). This remote area has been mapped by the Australian Bureau of Mineral Resources and BP geologists and geophysicists. There is also data from a number of exploratory boreholes drilled for petroleum.

The stratigraphic sequence is summarized in Figure 7-9, the emphasis being on the lithologies because these are the materials that are involved in the sliding. The Southern Highlands basin sequence begins with a transgressive Jurassic sequence of sands and mudstones (the Maril Shale) on granitic basement, followed by lower Cretaceous mudstones, siltstones (Kerabi Formation) and upper Cretaceous mudstones (Chim Formation). Unconformably above these are Paleocene mudstones and sandstones, Eocene limestone (deep-water micrites), and a lower Miocene shallow-water shelf limestone in the southern part, deep-water greywackes and mudstones in the northern part. In the thickest part of the basin, the Eocene limestone, which is about 300 m thick, overlies a sedimentary sequence at least 7,000 m thick that is composed largely of mudstone – in particular, the Chim Formation with about 3,000 m of mudstone, and the Maril Shale with about 1,500 m.

Mapping has revealed a deformation that can be conveniently divided into four zones, each with a characteristic style (Fig. 7-8). The northern zone (Zone IV of Jenkins) consists of a broad syncline on the southern slopes of the Kubor Range, with little obvious deformation. The next zone to the south, Zone III, is characterized by broad gentle synclines separated by narrow steep anticlines with sinuous trends. Some anticlines are overturned and overthrust to the south or south-west, and Jenkins notes a diapiric character in places. The next is a narrow zone (17 km) of imbricate thrusting in the Miocene limestones. The southernmost zone consists of large folds overthrust to the south or south-west over the uplifted Muller Range.

While this structural configuration may suggest sliding to the surface geologist, more positive evidence has been obtained from boreholes drilled for petroleum. Several cross-section through structures that have been drilled can be found in Jenkins (1974) and Ridd (1976). We take as an example the Puri anticline (Australasian Petroleum Co., 1961, p. 121, fig. 20), in the southernmost zone (I).

The well penetrated 616 m (2,022 ft) of Eocene and middle Miocene limestones (Fig. 7-10), then 1,646 m (5,403 ft) of Lower Cretaceous mudstone and siltstone before passing into Eocene and middle Miocene limestones again. The well terminated at 3,078 m (10,100 ft) in Lower Cretaceous mudstones. The overthrust was found by two side-tracked holes to be of very low angle of dip. The Mananda

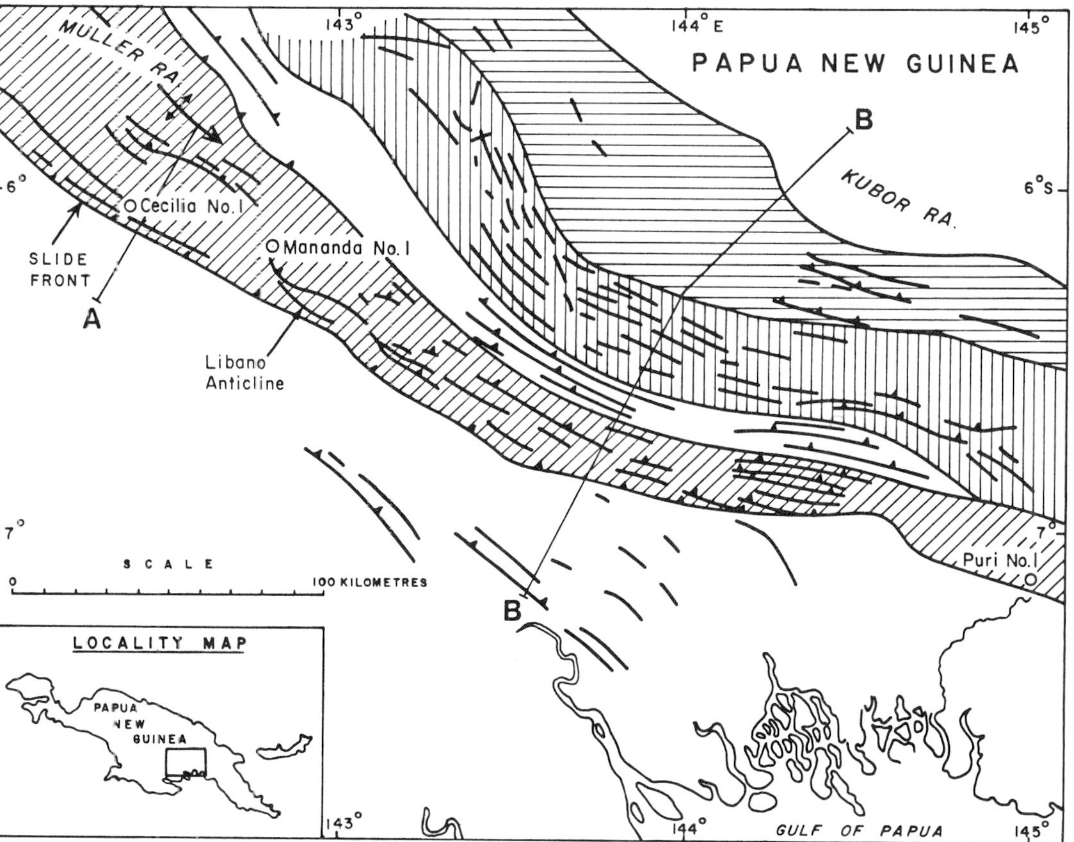

Fig. 7-8a. Structural map of western Papua New Guinea. (Courtesy of the British Petroleum Co., Ltd.) Structural Zone I, diagonal ornament: Zone II, plain (north of I): Zone III, vertical: Zone IV, horizontal ornament.

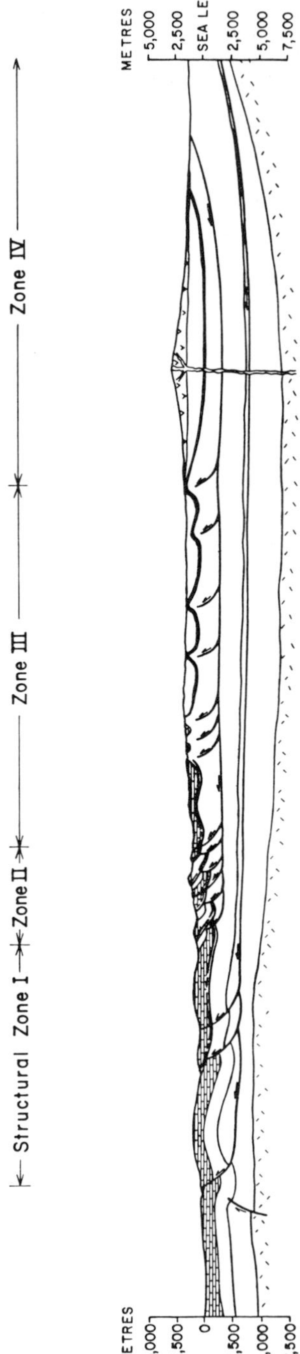

Fig. 7-8b. Section B-B of Figure 7-8a.

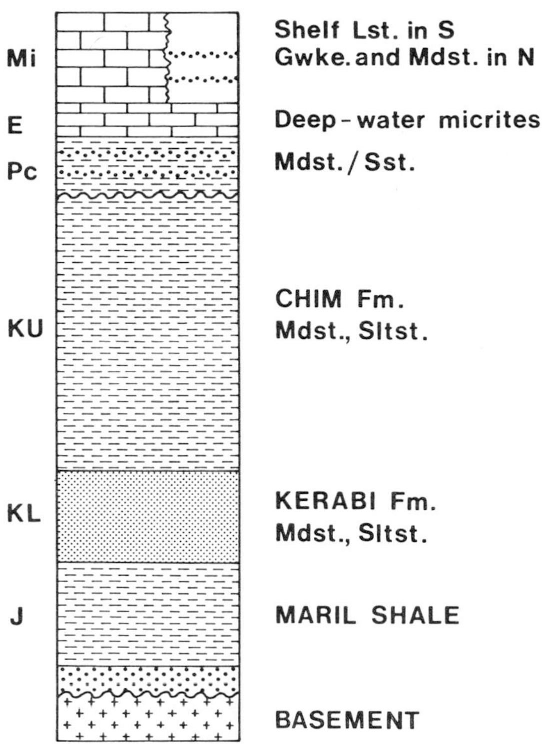

Fig. 7-9. Generalized stratigraphic sequence in area of sliding.

Cecilia, and Libano anticlines have very similar structural styles, also in Zone I but about 250 km to the west-north-west.

Uplift perhaps of the order of 10,000 m is indicated for the Kubor Range during the Plio-Pleistocene. The range stands now over 4,000 m above sea-level, with a Miocene turbidite sequence in outcrop both to the south and the north. A smaller uplift of 4,000 to 5,000 m is indicated for the Muller Range to the west, the axis of which lies about 130 km to the south-west of the Kubor Range. Between these two uplifts, basement reaches a depth of about 6,000 m, and this basement to the basin changes elevation by approximately 11 km over a distance of nearly 50 km – a slope of 12° to 13° – up to the Kubor Range on the north side of the basin. It is therefore natural that this large area of structural complexity should be interpreted as due to sliding towards the south down the slope as it was created. Jenkins (1974) suggested two surfaces of detachment (sliding), the deeper one in the upper Jurassic sequence leading to the thrusts of the southernmost structural zone, the other within the Cretaceous sequence leading to the thrusts of the other three northern zones. Findlay (1974), who studied an area adjacent to the east of Jenkins', postulated a single detachment within the Jurassic sequence.

There is little data concerning the pore-fluid pressures in the sequence; and it is

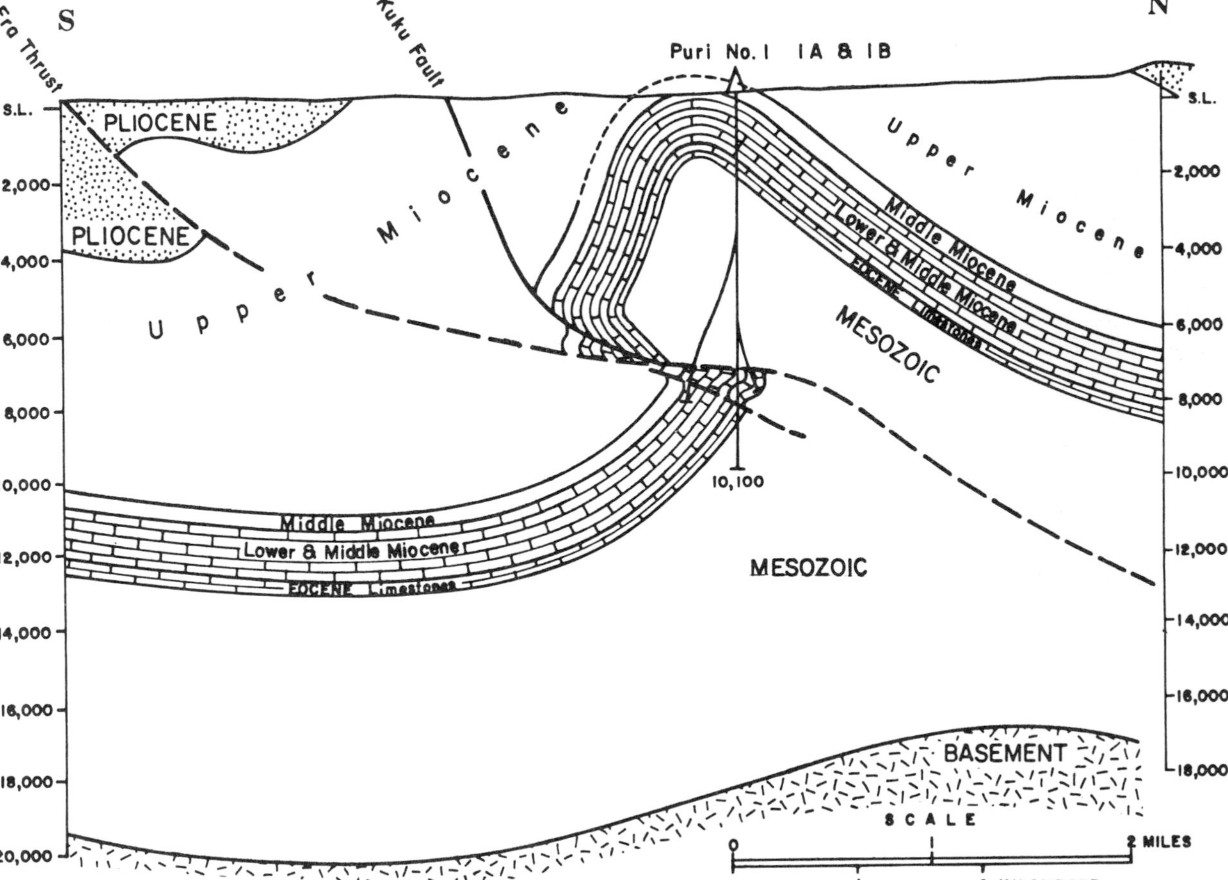

Fig. 7-10. Cross-section through the Puri anticline (depth in feet). (Courtesy of the British Petroleum Co., Ltd.)

unwise to infer pressures from the drilling difficulties that were sometimes encountered, because some of these were related to mudlosses in the limestone sections. Nevertheless, Cretaceous and Tertiary mudstones in the Aure through, close to the east, are abnormally pressured (Lepine and White, 1973), and it seems reasonable to suppose that the thicker mudstones in the Southern Highlands basin are, or were, abnormally pressured to some extent. And the deformation itself may have led to higher pore pressures.

An understanding of this sliding must rest on an understanding of the physical properties of the sedimentary sequence, and the slopes generated by subsidence and uplift. The key properties are those of the dense limestones lying on Mesozoic mudstones that we regard as lubricants for the sliding. While it may be misleading to think of detachment occurring on thin surfaces like a fault, there is no difficulty in regarding the Chim Formation as the upper *zone* of detachment. As regards the limestone overburden, there is probably a quantitative distinction to be made between the northern area with a thin Eocene limestone overlain by a Miocene turbidite sequence, and the southern area of thicker Miocene limestones. We note that this facies boundary is also close to the boundary between the diapiric-type deformation to the north and the imbricate overthrusts to the south. As we have seen, the Plio-Pleistocene uplift of the Kubor Range produced a slope of 12° or 13° to the south; but southerly sliding was opposed by the smaller uplift along the Muller Range axis. The zone of imbricate thrusting is on the northern side of this axis, and thus appears to be related to fracture-failure of the more competent limestones between these opposing slopes and their sliding forces. The southernmost structural zone contains relatively undeformed limestones around the Muller Range axis (Zone I of Jenkins).

If a block that is relatively rigid rests on a layer of lubricant, and there is no barrier to sliding, the rate of sliding is a function of the weight of the block, the angle of slope, the equivalent viscosity of the lubricant and its thickness. If all the dimensions and parameters remain constant, there will be no deforming stresses within the block, and the block will slide without folding or faulting. If, however, there is an obstacle to sliding, such as a reversal of slope, then the forces opposing sliding create a stress field within the block in which the component of lateral compression along the block increases in the direction of dip. When this stress exceeds the strength of the material, a thrust will develop, and the block will slide.

Such intuitive approaches to the problem of sliding lead to no great difficulties when there is an obvious barrier to sliding: but in the case under discussion, there is a more subtle influence. The weight of the block is its weight in the ambient fluid – air if it is subaerial, water if submarine. Plio-Pleistocene uplift has elevated most of the Eocene and Miocene marine limestones from below sea-level to above sea-level in the area of structural complexity. The bulk weight of water-saturated limestone above sea-level is about 25,000 N m^{-3}, but only about 15,000 N m^{-3} below sea-level (2,500 and 1,500 kg weight per cubic metre). So a uniform block of limestone resting on a uniform layer of lubricant, inclined so that the block is above sea-level at one end and below it at the other, does not have an uniform sliding potential. Unobstructed, a subaerial block will slide faster than an identical submarine block on the same slope and lubricant, because the subaerial driving force is greater than the submarine.

There is little difficulty in understanding the 'diapiric' structural zone qualitatively, with broad synclines separated by narrow anticlines or narrow belts of disturbance. This has the same character as the pre-orogenic deformation discussed earlier in this chapter, and a density inversion almost certainly exists between the deep-water micrites and underlying Chim Formation. The writer had this type of deformation in mind when he postulated that the dominant wavelength of diapirism could determine the length of thrust sheets (in the direction of movement) in unstable sedimentary sequences, the trend of such diapiric anticlines being parallel to the depositional strike, normal to the slope (Chapman, 1974). Each line of diapiric anticline forms a line of weakness, and any down-slope resistance to sliding will tend to deform the diapiric anticline into asymmetry and overthrusting.

The zone of imbricate thrusting south of the 'diapiric' zone may also have had a diapiric influence; but its position near the foot of the subaerial slope and the opposing slope of the Muller Range axis suggests that this deformation is due to fracture-failure in an area of high lateral compressive stress.

The sliding of the Tertiary, and perhaps Mesozoic, section in the Southern Highlands basin is considered to be an example of orogenic deformation that is gravity-induced, and, at least in part, superimposed on pre-orogenic deformation due to mechanical instability in the sequence. In these processes, the pore water almost certainly had the effect of reducing the equivalent viscosity of the thicker mudstones. The mudstones were evidently of sufficiently low viscosity for diapiric anticlines to form; and also the strain induced in the mudstones would have the effect of reducing bulk volume, or increasing the pore pressure if the pore water could not escape fast enough.

The role of sea-level in the style of deformation is a teasing problem of some interest. In the East Borneo basin of Kalimantan, in the region of the Mahakam delta (Fig. 7-11), the structural style onshore differs significantly from that found offshore. Onshore, the style is very similar to that of western Papua New Guinea, with long, sinuous, steep anticlines separated by broad, gentle synclines. The anticlinal trends tend to be asymmetrical, near the coast towards the coast, some with thrusts in the core. These anticlines have long been regarded as due to lateral compression: they formed an important part of van Bemmelen's *gravitational tectogenesis* (van Bemmelen, 1949, pp. 352, 732). However, drilling in the Mahakam delta and offshore from it has revealed growth anticlines with normal growth faults in sediments of similar (but rather younger, perhaps) age as those deformed onshore, from which we infer a stress field with a component of horizontal *tension* (Chapman, 1977a). The stratigraphic sequence is regressive, with a thick shale that is abnormally

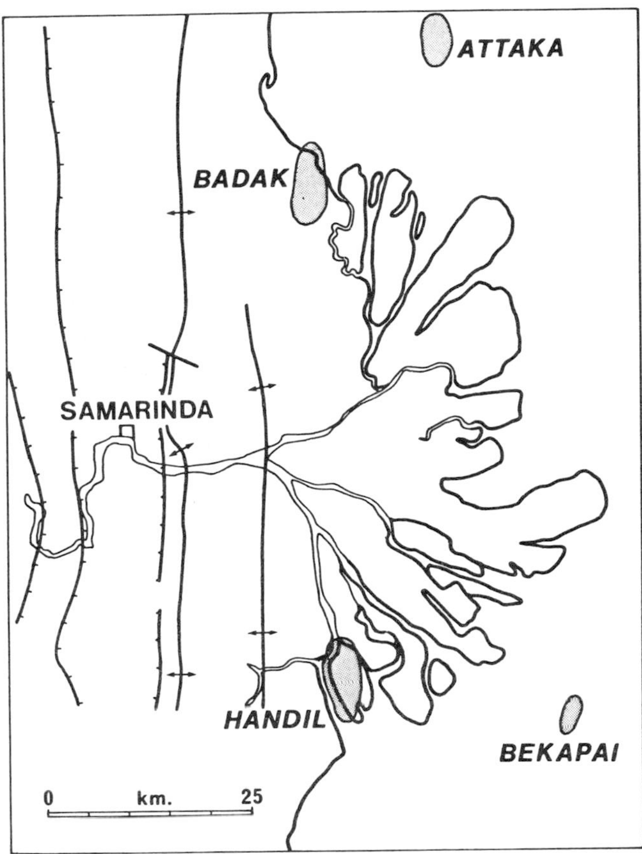

Fig. 7-11. Sketch map of Mahakam delta, Kalimantan, Indonesia (after Chapman, 1977a, fig. 5).

pressured (at very shallow depth in the cores of the anticlines, generally). When a map has been published of the contours on the top of abnormal pressures, we shall perhaps be able to understand this paradox. It seems likely that the onshore anticlines are diapiric (see Weeda, 1958, for example), and have been squeezed by the sliding of the sandy overburden on abnormally-pressured shales against the resistance offered by the submarine part of the sequence down-slope. Conversely, in the submarine part, if there is an increasing sliding potential seawards, the overburden may have a component of lateral tension.

In New Zealand, in the East Coast basin of North Island, Upper Cretaceous, Eocene, and Miocene sedimentary rocks are found with great structural complexity of the same general type as that found in Kalimantan and western Papua New Guinea – broad gentle synclines separated by steep narrow anticlines, with thrusts, overfolds, transcurrent faults, and mudvolcanoes (Ridd, 1970; Laing, 1972). Ridd attributed this structural style to sliding on the abnormally-pressured shales of early

Tertiary age that have been encountered in boreholes drilled for petroleum. Abnormally-pressured shales occur as shallow as 356 m (1,168 ft) in Rotokautuku 1 and the well had to be abandoned at 627 m (2,057 ft) on account of the difficulties (Katz, 1974, p. 469). It will be interesting to see if the offshore structural style is similar.

Finally, it must be noted that sliding is not necessarily towards the sea: its direction depends on the slopes generated. In the North Coast basin of West Irian, under the Mamberamo delta, drilling has not only revealed very shallow abnormally-pressured shales in a regressive sequence, but also that the basement is very shallow near the present-day coastline (Visser and Hermes, 1962, Enclosure I-III). Here it appears that the tendency is to slide towards the land; but there is too little data to pursure this further.

Mechanical problems of such complexity cannot reliably be understood by intuitive argument, so we shall seek a better insight into the role of pore water in such deformation by more rigorous mechanical analysis (although necessarily simplified).

SELECTED BIBLIOGRAPHY

Australasian Petroleum Co. Pty. Ltd., 1961. Geological results of petroleum exploration in western Papua 1937-1961. *J. Geol. Soc. Australia*, 8 (1): 1-133.

Barton, D.C., 1933. Mechanics of formation of salt domes with special reference to Gulf Coast salt domes of Texas and Louisiana. *Bull. American Ass. Petroleum Geologists*, 17 (9): 1025-1083.

Biot, M.A., and Odé, H., 1965. Theory of gravity instability with variable overburden and compaction. *Geophysics*, 30: 213-227.

Bishop, R.S., 1978. Mechanism for emplacement of piercement diapirs. *Bull. American Ass. Petroleum Geologists*, 62 (9): 1561- 1583.

Böttcher, H., 1925. Die Tektonik der Bochumer Mulde zwischen Dortmund und Bochum und das Problem der westfälischen Karbonfaltung. *Glückauf: Berg- und Hüttenmännische Zeitschrift*, 61: 1145- 1153, 1189-1194.

Böttcher, H., 1927. Faltungsformen und primare Diskordanzen im niederrheinisch-westfälischen Steinkohlengebirge. *Glückauf: Berg- und Hüttenmännische Zeitschrift*, 63: 113-121.

Bruce, C.H., 1973. Pressured shale and related sediment deformation: mechanism for development of regional contemporaneous faults. *Bull. American Ass. Petroleum Geologists*, 57 (5): 878-886.

Chapman, R.E., 1973. *Petroleum geology: a concise study*. Elsevier Scientific Publ. Co., Amsterdam, London, and New York, 304 pp.

Chapman, R.E., 1974. Clay diapirism and overthrust faulting. *Bull. Geol. Soc. America*, 85 (10): 1597-1602.

Chapman, R.E., 1976. *Petroleum geology: a concise study* (paperback edition). Elsevier Scientific Publ. Co., Amsterdam, Oxford, and New York, 302 pp.

Chapman, R.E., 1977a. Subsidence and deformation of terminal regressive sequences in the Indonesian region. *Proc. Indonesian Petroleum Ass.*, 5 (1) (for 1976): 151-158.

Chapman, R.E., 1977b. Petroleum geology of young regressive sequences. *Proc. S.E. Asia Petroleum Exploration Soc.*, 3 (for 1976): 8-38.

Crans, W., Mandl, G., and Haremboure, J., 1980. On the theory of growth faulting: a geomechanical delta model based on gravity sliding. *J. Petroleum Geology*, 2 (3): 265-307.

Dailly, G.C., 1976. A possible mechanism relating progradation, growth faulting, clay diapirism and overthrusting in a regressive sequence of sediments. *Bull. Canadian Petroleum Geology*, 24 (1): 92-116.

Daneš, Z.F., 1964. Mathematical formulation of salt-dome dynamics. *Geophysics*, 29 (3): 414-424.

Dennis, J.G., 1967. International tectonic dictionary English terminology. *Memoir American Ass. Petroleum Geologists*, 7 (196 pp.).

Dickinson, G., 1953. Geological aspects of abnormal reservoir pressures in Gulf Coast Louisiana. *Bull. American Ass. Petroleum Geologists*, 37 (2): 410-432.

Dron, R.W., 1900. The probable duration of the Scottish coalfields. *Trans. Instn Mining Engineers*, 18: 194-211. (Discussion: 211-212.)

Findlay, A.L., 1974. The structure of foothills south of the Kubor Range, Papua New Guinea. *J. Australian Petroleum Exploration Ass.*, 14 (1): 14-20.

Fowler, W.A., Boyd, W.A., Marshall, S.W., and Myers, R.L., 1971. Abnormal pressures in Midland Field, Louisiana. *In:* Houston Geol. Soc., *Abnormal subsurface pressure: a Study Group report* 1969-1971. Houston Geol. Soc., Houston, pp. 48-77.

Gansser, A., 1960. Über Schlammvulkane und Salzdome. *Vierteljahrsschrift der Naturforschenden Gesellschaft in Zürich*, 105: 1-46.

Harkins, K.L., and Baugher, J.W., 1969. Geological significance of abnormal formation pressures. *J. Petroleum Technology*, 21: 961-966.

Houston Geological Society, 1971. *Abnormal subsurface pressure: a Study Group report* 1969-1971. Houston Geol. Soc., Houston, 92 pp.

Hubbert, M.K., 1951. Mechanical basis for certain familiar geologic structures. *Bull. Geol. Soc. America*, 62: 355-372.

Hubbert, M.K., 1972. *Structural geology*. Hafner, New York, 329 pp. (*Contains Hubbert's papers on this theme.*)

Hubbert, M.K., and Rubey, 1959. Role of fluid pressure in mechanics of overthrust faulting, I. Mechanics of fluid-filled porous solids and its application to overthrust faulting. *Bull. Geol. Soc. America*, 70 (2): 115-166.

Hubbert, M.K., and Willis, D.G., 1957. Mechanics of hydraulic fracturing. *Trans. American Inst. Mining Metallurgical Petroleum Engineers*, 210: 153-166. (Discussion: 167-168.)

Jenkins, D.A.L., 1974. Detachment tectonics in western Papua New Guinea. *Bull. Geol. Soc. America*, 85 (4): 533-548.

Katz, H.R., 1974. Recent exploration for oil and gas. *In*: G.J. Williams (Ed.), Economic geology of New Zealand. *Monograph Series Australasian Inst. Mining Metallurgy*, 4: 463-480.

Kehle, R.O., 1970. Analysis of gravity sliding and orogenic translation. *Bull. Geol. Soc. America*, 81 (6): 1641-1664.

Laing, A.C.M., 1972. Geology and petroleum prospects of Ruatoria area, east coast, North Island, New Zealand. *J. Australian Petroleum Exploration Ass.*, 12 (1). 45-52.

Lepine, F.H., and White, J.A.W., 1973. Drilling in overpressured formations in Australia and Papua New Guinea. *J. Australian Petroleum Exploration Ass.*, 13 (1): 157-161.

Liechti, P., Roe, F.W., and Haile, N.S., 1960. The geology of Sarawak, Brunei and the western part of North Borneo. *Bull. Geol. Surv. Dept. British Territories Borneo*, 3.

Lyell, C., 1867. *Principles of Geology or the modern changes of the Earth and its inhabitants* (10th edition) *volume* 1. John Murray, London, 671 pp.

Morgan, J.P., Coleman, J.M., and Cagliano, S.M., 1968. Mudlumps: diapiric structures in Mississippi delta sediments. *In*: J. Braunstein and G.D. O'Brien (Eds), Diapirs and diapirism. *Memoir American Ass. Petroleum Geologists*, 8: 145-161.

Murray, G.E., 1961. *Geology of the Atlantic and Gulf Coastal province of North America.* Harper & Brothers, New York, 692 pp.

Parker, T.J., and McDowell, A.N., 1955. Model studies of salt-dome tectonics. *Bull. American Ass. Petroleum Geologists*, 39 (12): 2384-2470.

Ramberg, H., 1967. *Gravity, deformation and the Earth's crust as studied by centrifuged models.* Academic Press, London and New York, 214 pp.

Ramberg, H., 1972. Theoretical models of density stratification and diapirism in the Earth. *J. Geophysical Research*, 77 (5): 877-889.

Ridd, M.F., 1970. Mud volcanoes in New Zealand. *Bull. American Ass. Petroleum Geologists*, 54 (4): 601-616.

Ridd, M.F., 1976. Papuan basin – on-shore. *In*: R.B. Leslie, H.J. Evans, and C.L. Knight (Eds.),

Economic geology of Australia and Papua New Guinea – 3. Petroleum. *Monograph Series Australasian Inst. Mining Metallurgy*, 7: 478-494.

Schaub, H.P., and Jackson, A., 1958. The northwestern oil basin of Borneo. *In*: L.G. Weeks (Ed.), *Habitat of oil*. American Ass. Petroleum Geologists, Tulsa, pp. 1330-1336.

Selig, F., 1965. A theoretical prediction of salt dome patterns. *Geophysics*, 30: 633-643.

Stutzer, O., 1930. Absinken, Sedimentation und Faltung – gleichzeitige vorgange in manchen Erdolgebieten (Abstract). *Geologische Rundschau*, 21: 141.

Tanner, W.F., and Williams, G.K., 1968. Model diapirs, plasticity, and tension. *In*: J. Braunstein and G.D. O'Brien (Eds), Diapirism and diapirs. *Memoir American Ass. Petroleum Geologists*, 8: 10-15.

Thorsen, C.E., 1963. Age of growth faulting in southeast Louisiana. *Trans. Gulf-Coast Ass. Geol. Socs*, 13: 103-110.

Tiddeman, R.H., 1890. On concurrent faulting and deposit in Carboniferous times in Craven, Yorkshire, with a note on Carboniferous reefs. *Report British Ass. Advancement Science* (Newcastle-upon-Tyne, 1889), pp. 600-603.

Van Bemmelen, R.W., 1949. *The geology of Indonesia. Vol. 1A General geology of Indonesia and adjacent archipelagoes*. Government Printing Office, The Hague, 732 pp.

Visser, W.A., and Hermes, J.J., 1962. Geological results of the exploration for oil in Netherlands New Guinea. *Verhandelingen van het Koninklijk Nederlandsch Genootschap voor Nederland en Koloniën*, Geologische Serie, 20: 1-265.

Weeda, J., 1958. Oil basin of East Borneo. *In*: L.G. Weeks (Ed.), *Habitat of oil*. American Ass. Petroleum Geologists, Tulsa, pp. 1337-1346.

8. PORE WATER AND SLIDING

Large-scale sliding of geological sequences, with little internal deformation, is not a new idea. During the second half of the 19th Century, as the geology of the Alps, north-west Scotland, and Scandinavia was being unravelled, evidence emerged of lateral displacements of blocks many tens of kilometres long* in the direction of movement. For example, Törnebohm (1896, p. 194) postulated movement of blocks at least 130 km long on Caledonian thrusts. The main difficulty in these ideas was in understanding the mechanics. The paradox was this: the strength of the rock limits the length of the block that can be pushed along a horizontal surface because, if the force applied to the end exceeds the strength of the material, the block will fail by internal shear at the end being pushed. The strength of rocks is quite inadequate to support the push required to move blocks longer than a few kilometres. On the other hand, if the block slides down a slope under the force of gravity, the previous difficulty is replaced by two others: the coefficient of friction of rock on rock suggests that an angle of about 30° would be required for gravitational sliding – and that also implies a vertical relief of about half the length of the block. The restrictions on relief limit the length of blocks that slide under gravity to a few kilometres.

Smoluchowski (1909) calculated that the strength of granite was far less than that needed for a 160-km (100-mile) block to be pushed, but continued

> 'Suppose a layer of plastic material, say pitch, interposed between the block and the underlying bed; or suppose the bed to be composed of such material: then the law of viscous liquid friction will come into play, instead of the friction of solids; therefore any force, however small, will succeed in moving the block. Its velocity may be small if the plasticity is small, but in geology we have plenty of time; there is no hurry.'.

More and more evidence of large-scale translations was collected from different parts of the world during the first half of the 20th Century. It is probably true to say that the consensus of opinion was for pushing these large blocks or sheets along overthrusts rather than sliding them down slopes, although Reyer (1888) had suggested sliding as a cause of folding (Fig. 8-1). Indeed, evidence of uphill move-

* The conventional terminology of thrust sheets, long established, is *length* along the mapped thrust, *breadth* or *width* at right angles to the mapped thrust. When discussing the movement of such sheets, it is a needless distraction to have to remember that breadth or width is the length in the direction of movement (we do not talk of a motor car's being 2 m long and 4 m wide). So throughout this chapter, *length* means length in the direction of movement.

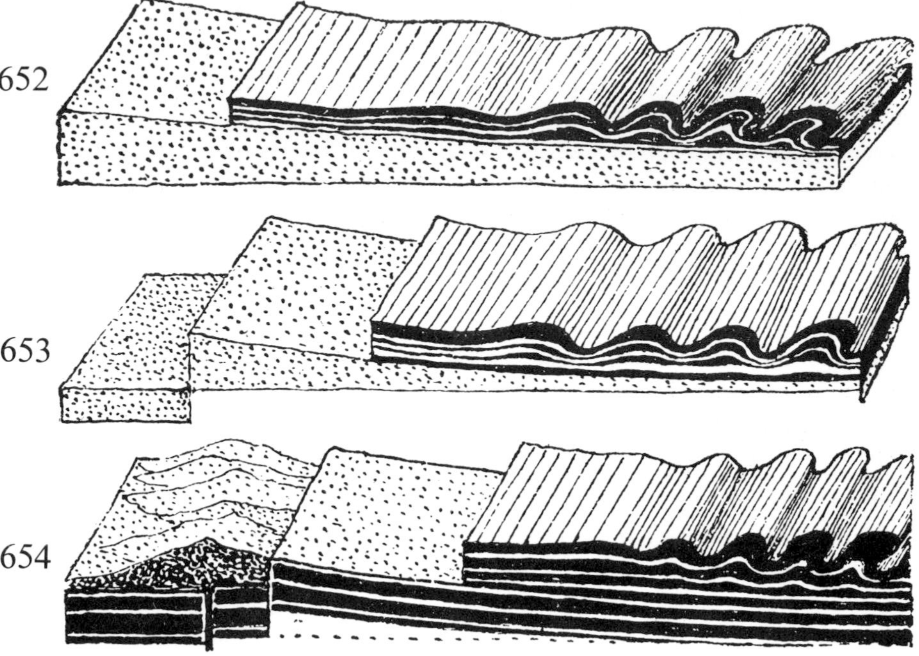

Fig. 8-1. Reyer's concept of sliding and folding (Reyer, 1888, figs. 652, 653, and 654).

ment in some areas, such as the Jura, apparently precluded gravitational sliding. The problems of gravitational sliding lay not so much in the evidence as the angle throught to be required.

The paradox appeared to be resolved by the fluid-pressure hypothesis of Hubbert and Rubey (1959), which was an extension of Terzaghi's hypothesis for landslides (Terzaghi, 1943, p. 235; 1950). In a theoretical analysis supported by experiment, they argued that pore pressure can relieve the effective normal stress to such an extent that longer blocks could be pushed without internal failure, and that blocks could slide down slopes much gentler than the angle indicated by the coefficient of friction.

This hypothesis has been widely accepted. However, it has usually been used out of context. The main part of Hubbert and Rubey's paper was concerned with *subaerial* sliding of water-saturated blocks; and in their application of the hypothesis to the overthrusts of western Wyoming (Rubey and Hubbert, 1959) they found it inadequate to account quantitatively for the inferred sliding, and so they concluded that the process of effective-stress relief *assisted* the sliding.

The difficulty in their hypothesis lies in the fact that their process is much more efficient subaerially than under water (by a factor of about two) and many slides appear to have been submarine. Indeed, it is hard to believe that blocks about 6 km thick in western Wyoming slid wholly subaerially, especially if they were sliding down the flank of a geosyncline (Rubey and Hubbert, 1959, p. 194). The less favourable mechanics of submarine sliding reduces the contribution of effective-stress relief through pore pressure. Moreover, almost all the sediments known to have pore pressures high enough for low-angle sliding are below sea-level. The paradox seemded to be re-appearing. Chapman (1979) discussed these difficulties at some length, and concluded that catastrophic slides, such as the Grand Banks slide reported by Heezen and Drake (1964), could have occurred by the Terzaghi/Hubbert and Rubey process.

Greater promise of resolving the paradox lies in the point made by Smoluchowski (1909): lubrication. The most widespread class of abnormally-pressured sediments is the thick shale section of young regressive sequences, commonly at depths of 1 to 3 km below sea-level. These shales have the properties of a lubricant in geological time – low equivalent viscosity and adequate thickness.

In this chapter we shall take a semi-quantitative approach to sliding, using idealized models, in order to seek an insight into the processes, and the role of pore water. We shall apply these to the area of Papua New Guinea described at the end of the previous chapter.

GRAVITATIONAL SLIDING

Unlubricated

When a rectangular block of thickness h is placed on a surface so inclined that it does not slide (Fig. 8-2), its weight in the ambient fluid (water or air), per unit area of the base,

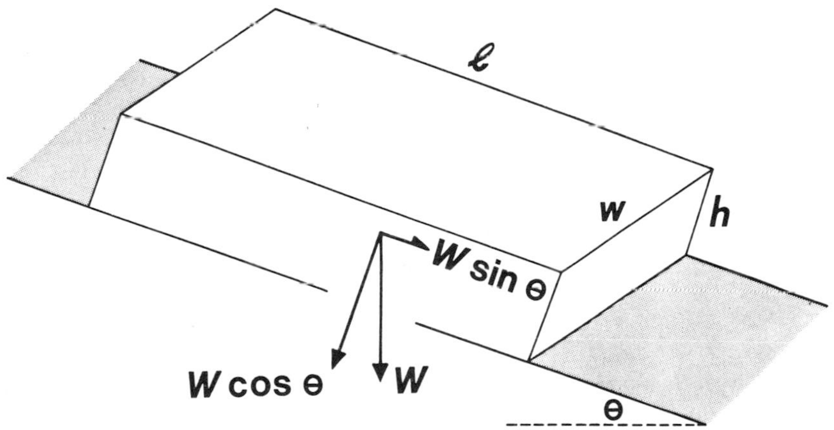

Fig. 8-2.

$$\sigma_z = \rho_b gh - \rho_a gh = (\rho_b - \rho_a) gh \tag{8.1}$$

can be resolved into a component normal to the surface

$$\sigma_n = (\rho_b - \rho_a) gh \cos\theta \tag{8.1a}$$

and a shear component

$$T = (\rho_b - \rho_a) gh \sin\theta. \tag{8.1b}$$

These may be regarded as *active* stresses; and they give rise to *reactive* stresses – a normal reaction, and frictional resistance (τ) that prevents sliding. As the angle of slope θ is increased, so the shear component of stress increases; and when this stress comes to exceed the frictional resistance, the block will slide. We shall assume that in the geological context, sliding will begin slowly when these two stresses are equal.

The Mohr-Coulomb criterion for simple unlubricated sliding (see Hubbert, 1951, p. 363; or any text-book on mechanics) is

$$\tau = \tau_0 + \sigma_n \tan\phi, \tag{8.2}$$

where τ_0 is the cohesive strength, or initial shear strength, of the materials at the surface that will become the sliding surface, when the normal stress σ_n is zero; $\tan\phi$ is the coefficient of sliding friction, ϕ commonly being about 30° for sedimentary rocks. Hence, sliding will take place when

$$T = \tau,$$
$$(\rho_b - \rho_a) gh \sin\theta = \tau_0 + (\rho_b - \rho_a) gh \cos\theta \tan\phi$$

(from equations (8.1a), (8.1b), (8.2)), or

$$\tan\theta = \frac{\tau_0}{(\rho_b - \rho_a) gh \cos\theta} + \tan\phi. \tag{8.3}$$

This equation indicates that sliding will usually take place at an angle θ rather greater than ϕ. It was Terzaghi (1943, 1950) who showed that for porous materials, the total normal stress is the sum of the effective stress and the pore-fluid pressure, $S = \sigma + p$, and that if $(S - p)$ is substituted for σ_n in equation (8.2),

$$\tau = \tau_0 + (S - p) \tan\phi, \tag{8.4}$$

pore-fluid pressure in a subaerial water-saturated block is seen to reduce the effective normal stress so that sliding may take place on a slope θ less than ϕ if τ_0 is sufficiently small.

Hubbert and Rubey (1959) argued that τ_0 is negligible in the geological context, and, defining a parameter $\lambda = p/S$ (the proportion of the total load supported by pore-fluid pressure) equated $T = S \tan\theta$ with $\tau = (1 - \lambda) S \tan\phi$, and so came to the expression

$$\tan\theta = (1 - \lambda) \tan\phi \tag{8.5}$$

for subaerial water-saturated blocks that may slide under gravity. From equation (8.5) it can be seen that as $p \rightarrow S$, so the angle at which sliding can occur approaches zero. Numerous *measured* values of $\lambda = 0.9$ indicate that sliding could take place on slopes as gentle as $3\frac{1}{2}^\circ$.

The difference between subaerial and submarine sliding relates to the effect of the ambient fluid: the subaerial block *in toto* receives negligible support due to buoyancy in air, but the submarine block receives considerable support due to buoyancy in water. Chapman (1979) derived the more general expression

$$\tan\theta = \frac{\tau_0}{(\rho_b - \rho_a)gh\cos\theta} + \delta\tan\phi, \tag{8.6}$$

where $\delta = (1 - \lambda)/(1 - \lambda_e)$, where λ_e is the proportion of the total load supported by the *ambient* fluid pressure. λ_e is sensibly zero when the ambient fluid is air, and equation (8.6) reduces to equation (8.5) when τ_0 is negligibly small. In the submarine environment, λ_e has the approximate value of 0.5; so the critical angle for submarine sliding is about twice that for subaerial sliding of an otherwise identical block. A water-saturated subaerial block sliding slowly down its critical angle of slope cannot slide far into the sea: the resistance to sliding remains the same, but the driving force is reduced from $\rho_b gh\sin\theta$ to $(\rho_b - \rho_a)gh\sin\theta$ (Fig. 8-3). If the initial shear strength cannot be ignored, the critical angle for sliding is increased, as Hsü (1969) pointed out. Table 8-1 shows the subaerial and submarine slopes required for blocks of various thicknesses when the initial shear strength is 3 MPa, and the slopes for zero initial shear strength, which are independent of thickness.

The fluid-pressure hypothesis seems inadequate as a general solution to the paradox of overthrust faulting and sliding, but it is entirely adequate as an expla-

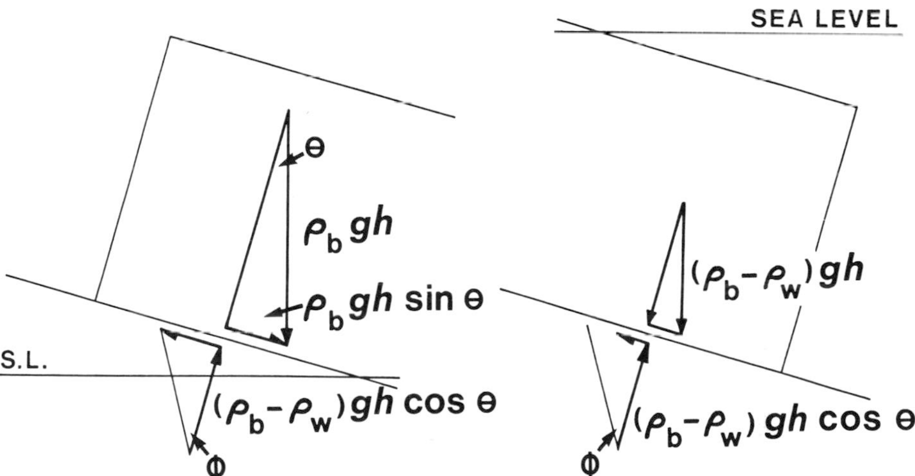

Fig. 8-3. When a sliding block enters the sea, the driving force is reduced but the frictional resistance remains the same.

Table 8-1. Critical angle of unlubricated gravitational sliding (degrees) for identical water-saturated subaerial and submarine blocks (*italics*). $\tau_0 = 3 \times 10^6$ Pa (30 bar).

Subaerial λ / *Submarine* δ	0.5 / *1.0*	0.6 / *0.8*	0.7 / *0.6*	0.8 / *0.4*	0.9 / *0.2*	1.0 / *0.0*
h km						
1	23	20	17	14	11	8
	42	*37*	*32*	*26*	*20*	*14*
2	20	17	14	10	7	4
	36	*31*	*26*	*20*	*13*	*7*
3	19	15	12	9	6	3
	34	*29*	*23*	*17*	*11*	*5*
4	18	15	12	8	5	2
	33	*28*	*22*	*16*	*10*	*3*
5	18	14	11	8	5	2
	32	*27*	*22*	*16*	*9*	*3*
$\tau_0 = 0$	16	13	10	7	3	0
	30	*25*	*19*	*13*	*7*	*0*

nation of some slides that have been discovered on the continental margins by marine seismic surveys. For example, the Grand Banks slide was attributed by Heezen and Drake (1964) to the 1929 earthquake, which led to a sequence of events well known to students of geology. The Grand Banks slide (Fig. 8-4) was about 50 km long (in the direction of movement), 400 m thick and lay on a slope of about 1°. We assume $\rho_b = 2{,}000$ kg m^{-3} and $\phi = 30°$. Re-arranging equation (8.6),

$$\tau_0 = (\tan\theta - \delta\tan\phi)(\rho_b - \rho_w)gh\cos\theta,$$

and inserting the above values, we find that τ_0 could not have been much greater than 7×10^4 Pa (0.7 bar), nor δ much greater than 0.03. In other words, the material at the sliding surface had very little cohesive strength, and the pore water was bearing almost the total overburden at the time of sliding.

These findings support the conclusion of Heezen and Drake. If the sliding surface were at the top of a poorly sorted, but porous, sand or silt, the seismic shock could have re-arranged the grains into a more stable packing. Such a change requires reduction of porosity – but porosity can only be reduced if pore water can be expelled. Until that happened, the pore water bore the overburden.

It is important to appreciate that the sliding surface is more likely to have been a sand or silt because good permeability is required for a short-distance slide: the excess pore pressures must dissipate quickly. Had the sliding surface been a clay, the slide would probably have travelled further because the excess pore-water pressure in clays cannot be dissipated quickly.

Chapman (1979) also developed expressions for the critical length of slide-blocks that are stopped by resistance at the down-slope end (a variation on the theme of

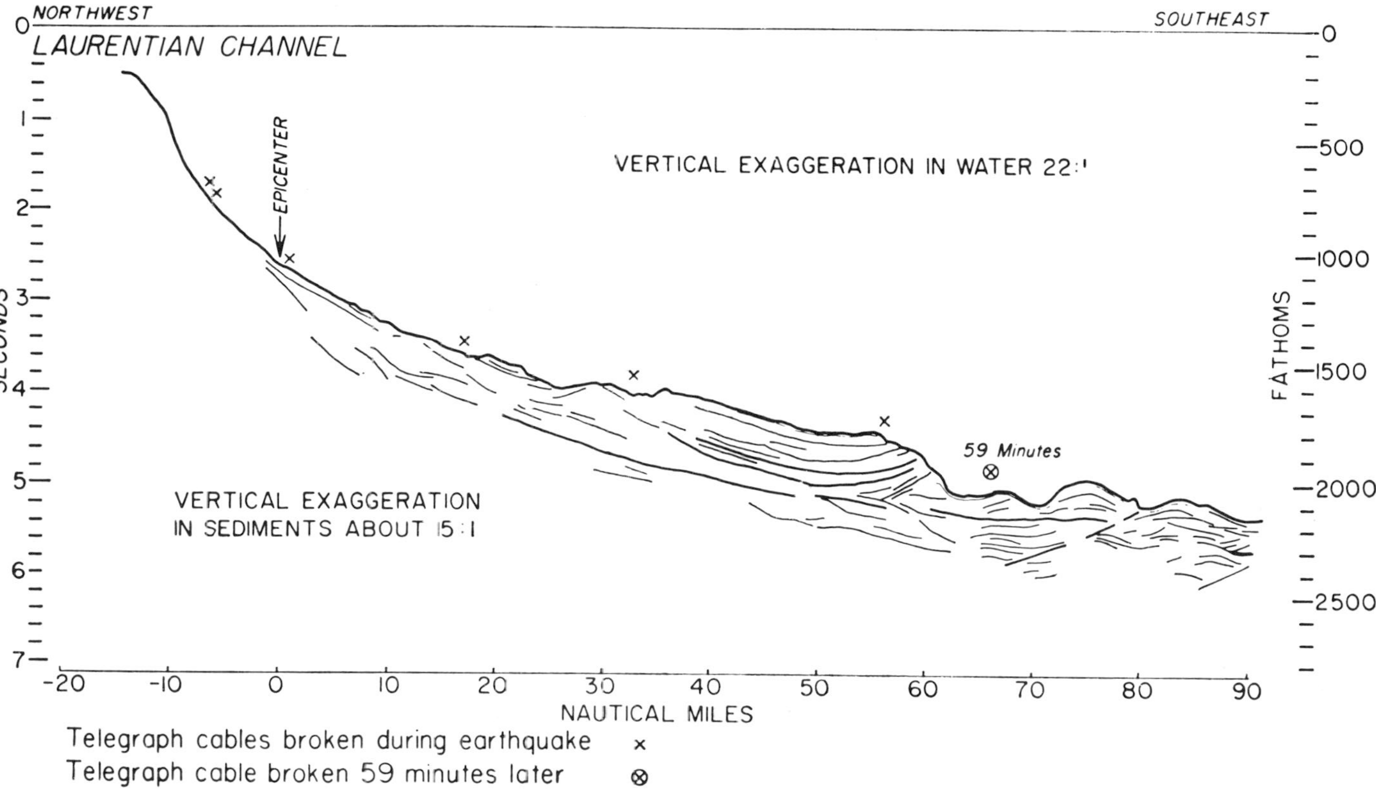

Fig. 8-4. Tracing of seismic reflection profile made on continental slope southeast of Laurentian Channel (Heezen and Drake, 1964, fig. 1).

pushing), and concluded that long blocks not only can but *must* slide on very small angles of slope. Steeper slopes have much shorter critical lengths and are likely to lead to chaotic slides and, in the extreme, turbidites. The length and slope of the Grand Banks slide are in good agreement with those predicted by theory.

This is possibly the mechanism by which turbidity currents and turbidites are caused. A sheet of sediment resting on a layer of water created by reduction of porosity in the underlying sand is momentarily lying on a very efficient lubricant. The expulsion of this lubricant implies a fluid-potential gradient directed towards the leading and trailing edges (at least; perhaps also laterally). Thus the excess pressure dissipates first at the margins of the slide. At the up-slope end, sediment will be stripped from the block. At the down-slope end, if the length is greater than the critical length (approximated in these conditions by $l_c = 3h/2\sin\theta$), the braking leading edge will not be able to sustain the shear component of weight of the block, and the slide will over-ride the leading edge. With material of little cohesive strength (a likely condition following a seismic shock), the block may then disintegrate, and the slide will continue as a turbidity current.

Lubricated sliding

If the block in Figure 8-2 is placed not on a rigid planar base but on a thickness of material with finite viscosity 'then the law of viscous liquid friction will come into play ... therefore any force, however small, will succeed in moving the block' (Smoluchowski, 1909). Let us examine this.

We must consider extensive sheets because a lubricant under a small block will be displaced with radial components, and the block will sink into the lubricant*. We follow Kehle (1970) in treating this as laminar *flow* down a slope.

An extensive thin sheet of a single liquid in uniform laminar flow down a gentle slope on a rigid planar surface (Fig. 8-5) can be regarded as obeying Newton's law of viscosity

$$\tau = \eta\, dV/dh, \tag{8.7}$$

where η is the dynamic, or absolute, viscosity. The units and dimensions of this relationship are

τ:	$\mathrm{Pa} = \mathrm{N\,m^{-2}} = \mathrm{kg\,m^{-1}\,s^{-2}}$	M	$L^{-1}\,T^{-2}$
dV:	$\mathrm{m\,s^{-1}}$		$L\ \ T^{-1}$
dh:	m		L
η:	$\mathrm{Pa\,s} = \mathrm{N\,s\,m^{-2}} = \mathrm{kg\,m^{-1}\,s^{-1}}$	M	$L^{-1}\,T^{-1}$

* This is indeed a real geological situation, but not the one we are considering. It has been examined experimentally by Ramberg (1967, p. 133). The sandy sequence of a delta may sink into its underlying clays or shales, and tend to extrude them. This seems to be happening in the Niger delta (Dailly, 1976).

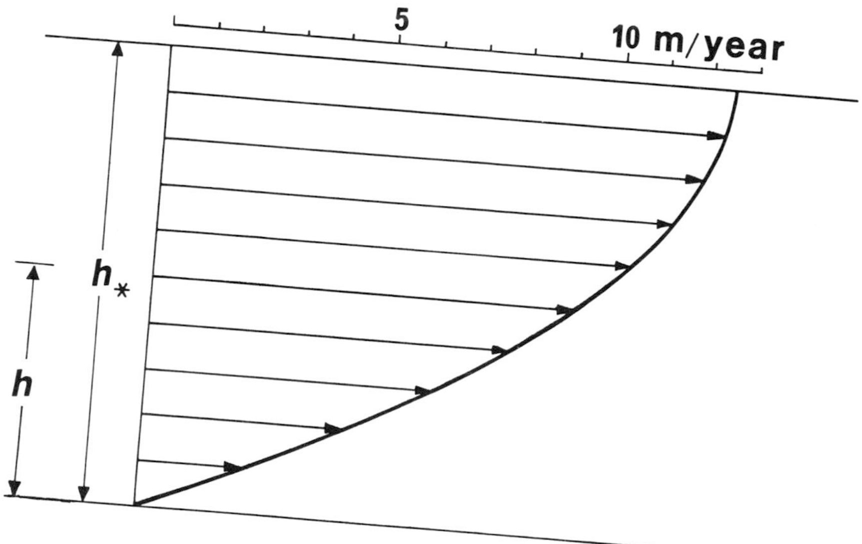

Fig. 8-5. Idealized velocity profile through a subaerial sheet of mudstone of mass density 2,300 kg m^{-3}, equivalent viscosity 10^{16} Pa s, 2 km thick, on slope of 5°.

Note that dynamic viscosity has the dimensions of *stress* × *length/velocity*, or *stress* × *time*. So a larger stress over a shorter time will have the same effect as a smaller stress over a longer time – a property used in the construction of properly scaled models.

Equivalent viscosity is a measure of the internal friction of the rocks at very small strain rates. In the geological context, with very large viscosities and very small velocities of sliding, we assume that

$$T = \tau - (\rho_b - \rho_a) g (h_* - h) \sin \theta, \tag{8.8}$$

where h_* is the total thickness of the material flowing, and $h_* - h$ is the thickness contributing to the shear stress at the level of interest (Fig. 8-5). Note that the shear stress is at a maximum on the rigid basal surface, where the velocity is zero, and zero at the top of the fluid where the velocity is greatest.

Equating equations (8.7) and (8.8),

$$\eta \frac{dV}{dh} = (\rho_b - \rho_a) g (h_* - h) \sin \theta,$$

and integrating this with respect to h (noting that $V = 0$ when $h = 0$) we obtain

$$V = \frac{(\rho_b - \rho_a)}{\eta} g \left(h_* h - \frac{h^2}{2}\right) \sin \theta. \tag{8.9}$$

Figure 8-5 shows the theoretical velocity profile through a subaerial shale of mass density 2,300 kg m^{-3}, assumed equivalent viscosity of 10^{16} Pa s, thickness 2 km, on a slope of 5°. Kehle found that the quantitative results are relatively insensitive to the flow law used, so we shall confine ourselves to the simple equation (8.9). But it must be assumed as a matter of course that the details of profiles constructed in this way are wrong: the velocity profile in a real shale will be different because the equivalent viscosity may well vary through the section. Our purpose is to understand the nature of lubricated sliding, and the order of magnitude of the difference of velocity across a lubricating bed. Equivalent viscosity can be regarded as the newtonian viscosity that would give the same effect as reality.

Consider now a rigid uniform bed one kilometre thick resting on one kilometer of the same shale with the same equivalent viscosity of 10^{16} Pa s on the same slope of 5° (Fig. 8-6). We shall assume a constant mass density of 2,300 kg m^{-3} because the velocity is relatively insensitive to small changes of mass density (the viscosity is the only parameter that can change the velocity by much more than an order of magnitude). The shear stress is the same at the top of the shale as in the middle of the 2-km shale considered earlier, and the velocity profile in our idealized model will be the same: the upper bed will slide over the rigid base at the velocity for $h_* = 2{,}000$ m, $h = 1{,}000$ m, that is, at about 3×10^{-7} m s^{-1}, or about 9 m/year.

The sequence and slope of Figure 8-6 are similar to those found in the area of sliding off the Kubor Range in Papua New Guinea (Jenkins, 1974). Jenkins estimated the distance slid to be about 13 km since the early Pliocene (approximately 6×10^6 years). Using this velocity, which is 7×10^{-11} m s^{-1} or 2×10^{-3} m/year, equation (8.9) implies an equivalent viscosity of the order of 10^{20} Pa s, or 10^{21} poise. This is within the range of viscosities inferred for the *mantle* (see Crittenden, 1963) – and the point should not be missed that significant translation is possible with such large viscosities.

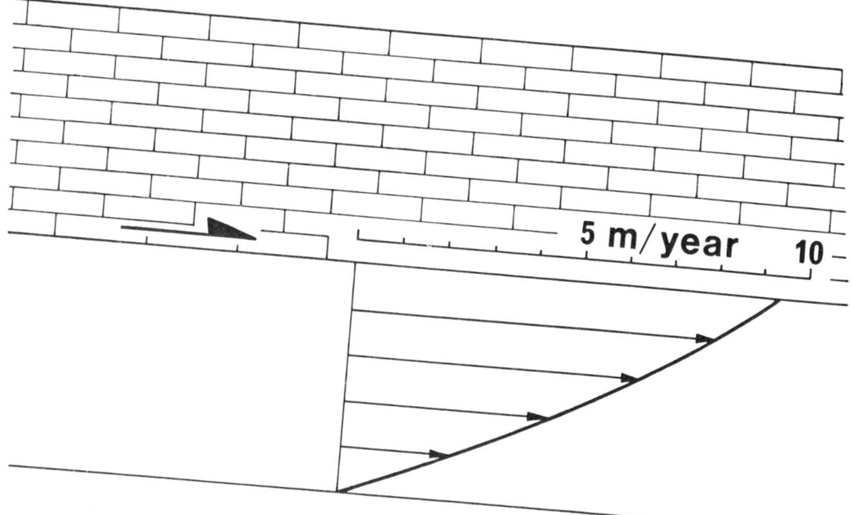

Fig. 8-6. Idealized velocity profile in 1 km of mudstone underlying 1 km of rigid material.

If we accept Kehle's figure of 10^{16} Pa s for the viscosity of shale, the velocity at the interface will be nearly 10 metres/year if the whole system is subaerial. This is, geologically, a catastrophic rate of sliding that would have completed the 13 km movement in about 1,500 years. Reducing the slope to 1° only reduces the velocity to 2 metres/year (7,000 years to complete the slide): reducing the thickness of the shale lubricant to 100 m on a 1° only reduces the velocity to 1/4 metre/year (54,000 years). If the whole system were submarine, these velocities would be about halved.

The idealized conditions assumed above only approach reality is so far as we can say that there exists a newtonian viscosity that would lead to the observed results. Qualitatively, the equivalent viscosity of a sedimentary rock depends on the effective stress, porosity, temperature, grain size and shape, and the material; it may also depend on the strain rate. It may therefore be very variable. When salt is the lubricant, some of these variables will be less important; but the viscosity of salt in the subsurface is unlikely to be much greater than 10^{16} Pa s, and may be as low as 10^{13} Pa s (see Odé, 1968, p. 74, Table IV).

The problem is seen to lie not so much in the answer to the question "How can blocks slide?" but rather "What slows them down?". Short answers to the second question are: a) surfaces are not planar, and non-planar surfaces consume energy, b) part of the sliding block, through fracture or otherwise, may come to slide over an unlubricated surface, and c) down-slope leads to the sea, where buoyancy roughly halves the driving force and halves the velocity.

The first two are rather outside the scope of this book, but let us look briefly at the question of unlubricated surfaces, because unlubricated sliding has been found to be inadequate for geological translation on a large scale. A combination of lubricated and unlubricated sliding may enlighten us.

Consider a block that slides slowly down a lubricated slope and meets an obstruction. If the block is short, it will merely stop at the obstruction; but there will be some critical length of the block which, if exceeded, will lead to failure at the obstruction in the form of a thrust fault (Fig. 8-7). The more the length exceeds the critical length, the greater the movement that will occur on the thrust fault. Such movement tends to lift the block off the thrust plane, but on geological scales the effect is to deflect the sliding surface upwards along the thrust plane, leading to an overthrust (Fig. 8-7b). As this movement continues, the frictional resisting forces on the unlubricated surface increase, while the driving force decreases. The slide will come to rest when these two forces are equal.

The driving force of an inclined block on a lubricated surface is simply the shear component of weight along the base of the block, that is

$$F_1 = l_1 w(\rho_b - \rho_a) gh \sin\theta, \tag{8.10}$$

where l_1 is the length of the block above the base of the thrust. The frictional resistance to sliding is given by the modified Mohr-Coulumb criterion,

$$F_2 = l_2 w \{\tau' + \delta(\rho_b - \rho_a) gh \cos\theta \tan\phi\}, \tag{8.11}$$

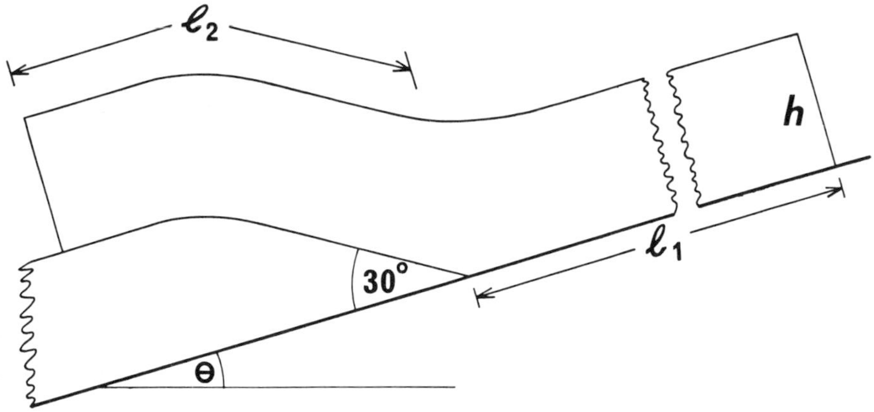

Fig. 8-7.

where l_2 is the length of the block below the base of the thrust, τ' is the shear stress that corresponds, during sliding, to the initial shear strength τ_0 before sliding (τ' is usually rather smaller than τ_0), and $\delta\,(\rho_b - \rho_a)\,gh\cos\theta$ is the effective normal stress (equations (8.4) and (8.6)) with the factor δ taking the pore-fluid and ambient-fluid pressures at the sliding surface over l_2 into account.

Sliding is also opposed by the shear component of weight down the thrust plane. The angle of the thrust will be about $(45^\circ - \phi/2)$ to the longitudinal greatest principal stress (which we take to be parallel to the base of the block), that is, about 30°. So we can take the length of the thrust plane to be about $2h$. This force is

$$F_3 = 2\,h\,w\,(\rho_b - \rho_a)\,gh\sin(30 - \theta). \tag{8.12}$$

The slide will therefore come to a halt when

$$F_1 = F_2 + F_3,$$

$$l_1 w\,(\rho_b - \rho_a)\,gh\sin\theta = l_2 w\,\{\tau' + \delta(\rho_b - \rho_a)\,gh\cos\theta\tan\phi\} + 2hw(\rho_b - \rho_a)\,gh\sin(30 - \theta)$$

from which

$$l_1 = l_2\left\{\frac{\tau'}{(\rho_b - \rho_a)\,gh\sin\theta} + \frac{\delta\cos\theta\tan\phi}{\sin\theta}\right\} + \frac{2h\sin(30 - \theta)}{\sin\theta}. \tag{8.13}$$

The minimum length of l_1, the maximum length of over-thrust, occurs when τ' is negligibly small compared to $(\rho_b - \rho_a)\,gh$, so, for fully submarine or fully sub-aerial slides, we can write for the maximum length of overthrust and thrust

$$l_2 \leqslant \frac{l_1\,sin\,\theta}{\delta\cos\theta\tan\phi} - \frac{2h\sin(30 - \theta)}{\delta\cos\theta\tan\phi}. \tag{8.14}$$

This is a linear relationship in l_1 and l_2 for constant slope and conditions, and gives the maximum likely length of an overthrust when the block length is $l = l_1 + l_2$ and the overthrust is unlubricated. Figure 8-8 shows the maximum length of the overthrust plotted as a percentage of the total length for blocks of various thicknesses on slopes of 10°.

If the weight of the block is to cause the thrust and overthrust, then the total length of the block above the obstruction must exceed its critical length (see Chapman, 1979, p. 26, eq. 30). But this is not the only possibility. The dominant wavelength of diapiric anticlines may provide natural lines of weakness and divide the block into lengths shorter than its critical length (Chapman, 1974).

Figure 8-9 shows the section between the Lavani anticline in the Muller Range and the Cecilia anticline in the area of Papua New Guinea studied by Jenkins (1974). The total length of this limestone block is about 30 km, 23 of which are on a slope of 10° above the base of the thrust (l_1). Roughly 1/3 of the block is subaerial. The limestone thickness averages ~ 1,250 m. Figure 8-10 is a plot of percentage overthrust against total length of block ($l_1 + l_2$) computed from equation (8.14). The maximum length of thrust and overthrust expected fom this theory is therefore 9 or 10 km (32%) for a fully subaerial slide, or 6 km (20%) for a fully submarine slide. This is in satisfactory agreement with the field observations of 7 km – and the slide may still be going on. In the upper part, there are about 7 km of the block above sea-

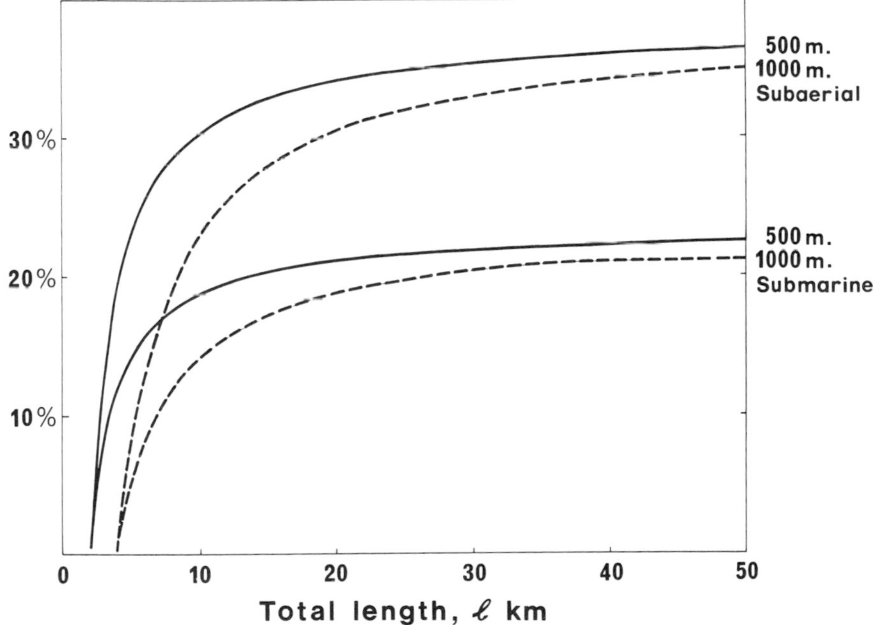

Fig. 8-8. Maximum length of thrust and overthrust expresses as a percentage of the total length of the block on slope of 10°.

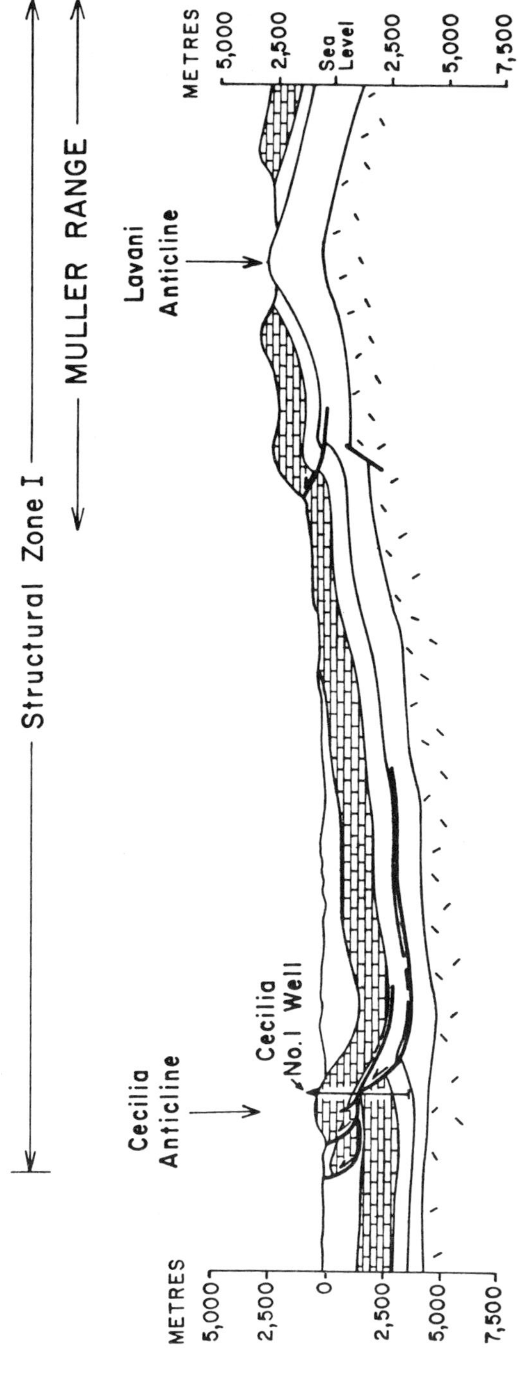

Fig. 8-9. Cross-section A-A of Figure 7-8a.

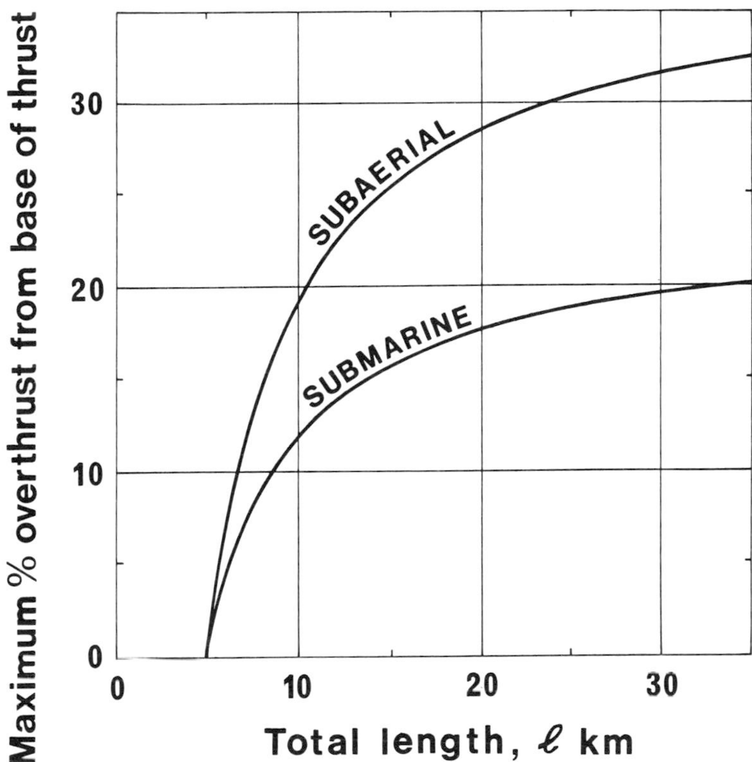

Fig. 8-10. Maximum length of thrust and overthrust for block 1250 m thick on slope of 10°, such as the limestone block in Figure 8-9.

level, above a small thrust fault. Here theory predicts an overthrust of less than 10%.

If we take Jenkins' (1974, p. 544) generalized dip of 4° off the Kubor Range at the base of the Tertiary, the maximum length of thrust and overthrust in a block 125 km long would be between 11 km (submarine) and 22 km according to theory if the average thickness of the block is between 1 and $1\frac{1}{2}$ km. This is also in satisfactory agreement with the field observation of 13 km.

It seems therefore reasonable to conclude that the reason why sliding has not been more extensive is that the obstructions to sliding have caused thrusting and overthrusting, part of which is unlubricated. We can therefore agree with all Jenkins' conclusions except that sliding is probably still going on.

EFFECT OF SEA-LEVEL

The effect of sea-level on sliding is of interest. Not only may blocks slide into the sea, but they may also be raised above sea-level during orogeny (as in the example from Papua New Guinea).

Consider a block sliding slowly down a constant slope on a lubricant of constant properties (Fig. 8-11). While it is entirely subaerial, the velocity of the block (equation (8.9)) is given by

$$V_{sa} = \frac{\rho_b g h \sin\theta}{\eta}\left(h_* h - \frac{h^2}{2}\right),$$

and when it is entirely submarine, its sliding velocity is given by

$$V_{sm} = \frac{(\rho_b - \rho_a) g h \sin\theta}{\eta}\left(h_* h - \frac{h^2}{2}\right).$$

Sea-level is a natural impediment to sliding because buoyancy reduces the shear component of weight, reducing the velocity of a lubricated slide by about half – and

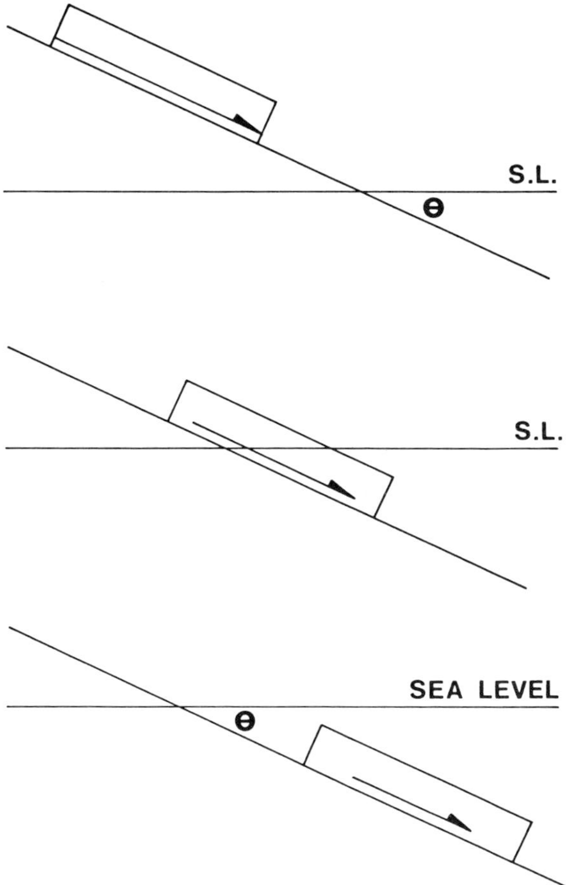

Fig. 8-11. Effect of sea-level on block sliding slowly down constant slope into sea. The lengths of the arrows are proportional to the component of weight down the slope.

normally stopping an unlubricated slide. As a lubricated block slides into the sea, a component of longitudinal compression is induced in the block. This increases to a maximum near sea-level when half the block is submerged. This force is the difference between the shear components of subaerial and submarine weights. It seems that this force alone will not cause significant deformation unless the sliding block is thin and weak.

Elevation of a block from below sea-level to above is another matter because, if equation (8.9) has any validity, blocks will slide off even gentle slopes much faster than geological rates of elevation. A sliding velocity of 10 m/year on a slope of 5° implies an elevation loss of nearly 1 m/year, which is several orders of magnitude greater than normal elevation rates (the Kubor Range rate of uplift has been of the order of 10^{-3} m/year). So, obstruction to sliding must exist from early on if a submarine block is to be elevated above sea-level. In Papua New Guinea, it seems essential for the preservation of the Tertiary sequence that the Muller Tange uplift was contemporaneous with the Kubor Range uplift.

If two contemporaneous uplifts retain a sequence that includes a lubricant, then the obstruction to sliding is the opposing slope, and the obstacle that we considered earlier is the part of the block on the opposing slope. In this case, there may be some up-slope sliding if the slopes are not in mechanical equilibrium. The component of longitudinal compression induced by these slopes (above or below sea-level) is greatest at the inflexion at the bottom of the slopes, and first failure is likely to occur there once some critical relief has been achieved (the critical relief being ($l_c \sin \theta$).

Once the overthrust requires for its maintenance of force greater than the strength of the block, another thrust will develop on top of the first. This may be repeated until the critical relief or length is no longer exceeded. It is tempting to speculate that this is the process by which overthrusts may be superimposed and imbricate structures developed.

SELECTED BIBLIOGRAPHY

Birch, F., 1961. Role of fluid pressure in mechanics of overthrust faulting: discussion. *Bull. Geol. Soc. America*, 72 (9): 1441-1444.

Chapman, R.E., 1974. Clay diapirism and overthrust faulting. *Bull. Geol. Soc. America*, 85 (10): 1597-1602.

Chapman, R.E., 1979. Mechanics of unlubricated sliding. *Bull. Geol. Soc. America*, 90: 19-28.

Chapple, W.M., 1978. Mechanics of thin-skinned fold-and-thrust belts. *Bull. Geol. Soc. America*, 89 (8): 1189-1198.

Crittenden, M.D., 1963. Effective viscosity of the Earth derived from isostatic loading of Pleistocene Lake Bonneville. *J. Geophysical Research*, 68 (19): 5517-5530.

Dailly, G.C., 1976. A possible mechanism relating progradation, growth faulting, clay diapirism and overthrusting in a regressive sequence of sediments. *Bull. Canadian Petroleum Geology*, 24 (1): 92-116.

De Jong, K.A., and Scholten, R., (Eds), 1973. *Gravity and tectonics*. John Wiley & Sons, New York, 502 pp.

Guterman, V.G., 1980. Model studies of gravitational gliding tectonics. *Tectonophysics*, 65: 111-126.

Heezen, B.C., and Drake, C.L., 1964. Grand Banks slump. *Bull. American Ass. Petroleum Geologists*, 48 (2): 221-225.

Hsü, K.J., 1969a. Role of cohesive strength in the mechanics of overthrust faulting and of landsliding. *Bull. Geol. Soc. America*, 80 (6): 927-952.

Hsü, K.J., 1969b. Role of cohesive strength in the mechanics of overthrust faulting and of landsliding: reply [to discussion by M.K. Hubbert and W.W. Rubey]. *Bull. Geol. Soc. America*, 80 (2): 955-960.

Hubbert, M.K., 1951. Mechanical basis for certain familiar geologic structures. *Bull. Geol. Soc. America*, 62: 355-372.

Hubbert, M.K., 1972. *Structural geology*. Hafner, New York, 329 pp.

Hubbert, M.K., and Rubey, W.W., 1959. Role of fluid pressure in mechanics of overthrust faulting, I. Mechanics of fluid-filled porous solids and its application to overthrust faulting. *Bull. Geol. Soc. America*, 70 (2): 115-166.

Hubbert, M.K., and Rubey, W.W., 1960. Role of fluid pressure in mechanics of overthrust faulting: a reply [to H.P. Laubscher]. *Bull. Geol. Soc. America*, 71 (5): 617-628.

Hubbert, M.K., and Rubey, W.W., 1961a. Role of fluid pressure in mechanics of overthrust faulting, I. Mechanics of fluid-filled porous solids and its application to overthrust faulting: reply to discussion by Francis Birch. *Bull. Geol. Soc. America*, 72 (9): 1445-1451.

Hubbert, M.K., and Rubey, W.W., 1961b. Role of fluid pressure in mechanics of overthrust faulting: a reply to discussion by Walter L. Moore. *Bull. Geol. Soc. America*, 72 (10): 1587-1594.

Hubbert, M.K., and Rubey, W.W., 1969. Role of cohesive strength in the mechanics of overthrust faulting and of landsliding: discussion. *Bull. Geol. Soc. America*, 80 (6): 953-954.

Jenkins, D.A.L., 1974. Detachment tectonics in western Papua New Guinea. *Bull. Geol. Soc. America*, 85 (4): 533-548.

Kehle, R.O., 1970. Analysis of gravity sliding and orogenic translation. *Bull. Geol. Soc. America*, 81 (6): 1641-1664.

Laubscher, H.P., 1960. Role of fluid pressure in mechanics of overthrust faulting: discussion. *Bull. Geol. Soc. America*, 71 (5): 611-616.

Moore, W.L., 1961. Role of fluid pressure in overthrust faulting: a discussion. *Bull. Geol. Soc. America*, 72 (10): 1581-1586.

Normark, W.R., 1974. Ranger submarine slide, northern Sebastian Vizcaino bay, Baja California, Mexico. *Bull. Geol. Soc. America*, 85: 781-784.

Odé, H., 1968. Review of mechanical properties of salt relating to salt-dome genesis. *In*: J. Braunstein and G.D. O'Brien (Eds), Diapirism and diapirs: a symposium. *Memoir American Ass. Petroleum Geologists*, 8: 53-78.

Price, N.J., 1977. Aspects of gravity tectonics and the development of listric faults. *J. Geol. Soc. London*, 133: 311-327.

Raleigh, C.B., and Griggs, D.T., 1963. Effect of the toe in the mechanics of overthrust faulting. *Bull. Geol. Soc. America*, 74: 819-830.

Ramberg, H., 1967. *Gravity, deformation and the Earth's crust as studied by centrifuged models*. Academic Press, London and New York, 214 pp.

Reyer, E., 1888. *Theoretische Geologie*. E. Schweizerbart'sche Verlaghandlung, Stuttgart, 867 pp.

Rubey, W.W., and Hubbert, M.K., 1959. Role of fluid pressure in mechanics of overthrust faulting, II. Overthrust belt in geosynclinal area of western Wyoming in light of fluid-pressure hypothesis. *Bull. Geol. Soc. America*, 70 (2): 167-206.

Smoluchowski, M.S., 1909. Some remarks on the mechanics of overthrusts. *Geological Magazine*, new series, decade V, v. 6: 204-205.

Terzaghi, K., 1943. *Theoretical soil mechanics*. Chapman & Hall, London; John Wiley & Sons, New York, 510 pp.

Terzaghi, K., 1950. Mechanism of landslides. *In*: S. Paige (Chairman), *Application of geology to engineering practice (Berkey Volume)* Geol. Soc. America, Boulder, pp. 83-124.

Törnebohm, A.E., 1896. Grunddragen af det centrala Skandinaviens bergbyggnad. *Kongliga Svenska Vetenskaps-Akademiens Handlingar*, 28 (5): 1-178. (Résumé in German: 179-197.)

9. CONCLUSION

We revert in conclusion to the central theme of this book – the movement of water in the subsurface. The principles we have developed and discussed can be applied to a range of geological problems either as qualitative or as semi-qualitative arguments. We take one of each for illustrative purposes.

FAULTS AND WATER MOVEMENT

An important contemporary debate amongst petroleum geologists is the question of faults' acting as conduits for subsurface fluids, particularly with regard to the migration of oil and gas, and abnormal pressures. It may seem at first sight that it is eminently plausible that fluids could be conducted upwards along faults (or downwards): but plausible arguments must be supported by physical principles if they are to be elevated even to the rank of pure speculation.

If a fault is to be a conduit for fluid *flow* (as distinct from some other mechanism), then it must have permeability, and there must be a potential gradient in the fault plane. If it is to be more important as a conduit than the adjacent sediments, the permeability must be better, or the potential gradient greater, in the plane of the fault than in the sediments. This implies a potential gradient between the fault plane and the adjacent sediments, and so requires a permeability barrier between the fault plane and the adjacent sediments.

It is evident that there are difficulties to be overcome if we are to accept faults as conduits for subsurface fluids in many geological contexts.

One possibility is that the fault plane develops void-channels, or channels of enhanced permeability, by separation of the walls. In such a case, the resistance to flow in these channels will be so much less than that through the sediments that a significant contribution could come from this cause. Normal faults are the most likely candidates for such a process in most areas, with transcurrent faults more likely in some. The stress field of a normal fault is such that the least principal stress is approximately horizontal, normal to the trace of the fault. But the least principal stress is *compressive* at all but relatively shallow depths (some hundreds of metres, not thousands, for sediments: Hubbert, 1951, p. 367; Hubbert and Willis, 1957), and soon exceeds the cohesive strength of the sediments. The tendency will be to close any gap quickly, if one can form at all.

A variation of this possibility is the development of a fault breccia that is more permeable than the adjacent sediments, and sealed from them to some extent by fault gouge. While the formation of breccia implies greater cohesive strength, it is unlikely to provide a conduit over a great depth-range.

More fundamentally, we have seen that the quantity of water that flows through permeable material is proportional to the area of cross-section normal to the flow. The cross-sectional area of a fault is infinitessimal compared to the area of adjacent sediment, and so the fault would have to have a vastly greater permeability or hydraulic gradient or both to move significant quantities of water upwards by this conduit.

It is very difficult to understand, with present knowledge, how a fault could possibly be an important conduit for subsurface fluids from depths greater than about one kilometre (to take a conservative round figure). The writer knows of two

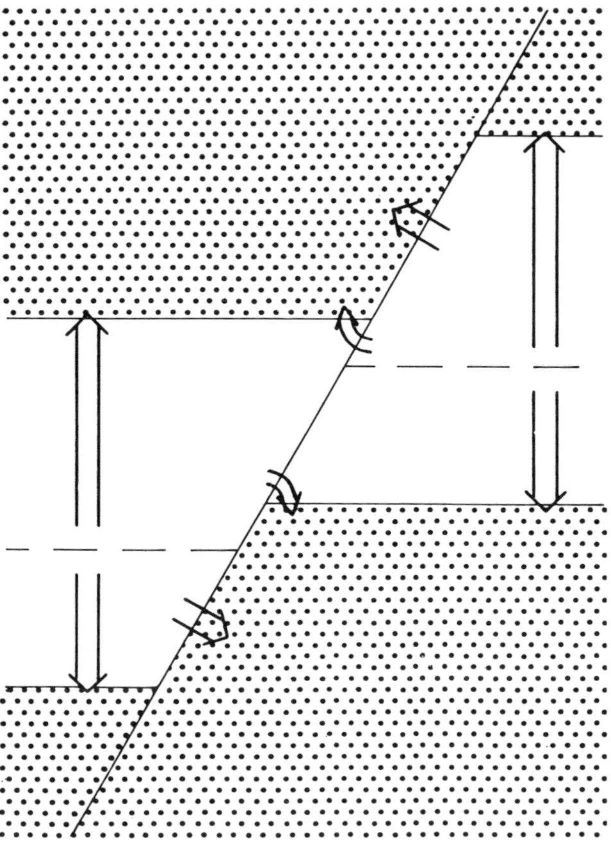

Fig. 9-1. Schematic section through fault cutting abnormally-pressured shale, showing pore-water flow directions.

oil-well blowouts that caused fluids to follow a fault to the surface from shallow depths, one of which repressured petroleum reservoirs at greater depths. Indeed, faulted oil fields would be rare, one imagines, if faults acted as conduits.

There are many examples of faults in petroleum fields acting as seals to the lateral flow of oil and gas.

Consider a fault penetrating a sequence in which there is a thick shale with pore pressures above normal hydrostatic (Fig. 9-1), the formations above and below having normal hydrostatic pressures. The potential gradients can be readily inferred, and the fault seems to play a part only in the lateral migration of fluids.

However, faults do not always dispace the abnormal pressures by the same amount as the sediments (see, for example, Fowler et al., 1971, p. 56, fig. 3). A situation such as that depicted in Figure 9-2 leads to a slightly different pattern of potential gradients, but there is still no obvious gradient up the faults.

It is important to understand that speculation devoid of physical support, or in seeming conflict with physical principles, may still turn out to be correct. But if we ignore our understanding of physical principles when speculating, we are more likely to be led into error. Temperature or geochemical anomalies near a fault must have explanations that are consistent with physical principles.

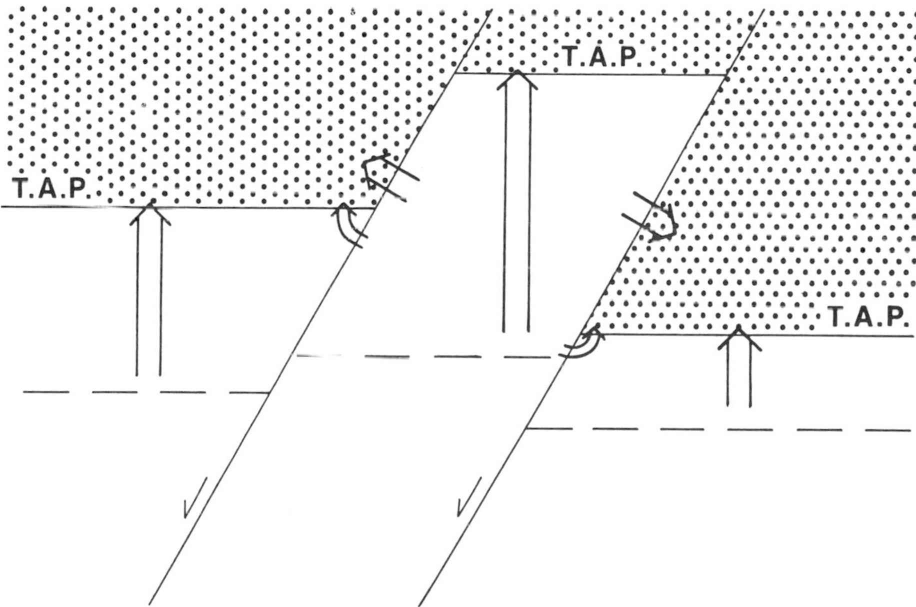

Fig. 9-2. Schematic section through faults cutting abnormally-pressured shales in which the degree and stratigraphic extent of the abnormality changes across the faults. Arrows show the direction of pore-water flow.

FLOW OF TWO IMMISCIBLE LIQUIDS IN POROUS SEDIMENTS

If analysis of the flow of a single liquid through porous materials has difficulties, that of flow of more than one liquid presents formidable problems. It is, of course, a matter of central interest to petroleum reservoir engineers, and much experimental and theoretical work has been done on it. But interest should not be confined to petroleum reservoir engineers because two-phase flow can occur in a variety of geological situations. We take a semi-quantitative approach because we are more interested in acquiring a feel for the parameters that affect the phenomenon than in developing predictive equations required for proper petroleum reservoir management. We shall take a simplistic approach along the lines of Chapter 3, and we shall regard both water and oil as incompressible liquids, evenly distributed in the sense that the proportion of either liquid in the total pore volume is considered to be present in each pore.

We cannot see two-phase flow in the subsurface, so its character must be inferred. When an oil well is put on production, oil typically flows with virtually no water, and it will produce a significant proportion of its total ultimate yield with very little water. Towards the end of the life of the well, the water content in the production increases – slowly at first, but accelerating – until the well is no longer economical; and when the well is abandoned, there is residual oil left in the reservoir amounting typically to about 65% of the pore space (Bridgeman, 1969). This suggests that the wetting liquid, water, is essentially static around the grains when the oil saturation of the pore spaces is high. As the water saturation increases, the reservoir rock becomes more permeable to water, while still retaining some permeability to oil. There is clearly an *effective permeability to oil* in the wetted reservoir rock that is analogous to intrinsic permeability.

Consider a porous, permeable, isotropic and homogeneous reservoir rock that is entirely saturated with water. We can determine from the flow parameters the intrinsic permeability of the material and its hydraulic conductivity. When oil occupies part of each pore (the material being water-wet), each globule of oil occupies the space of minimum potential in each pore – that is, the central part. As the globule grows, three distinct stages are recognizable:

1) the globules are smaller than the pore-throats connecting one pore with another, and there will be no great impediment to the flow of this oil with the water;
2) the globules are larger than the pore-throats, and work will have to be done to force the globule into the next pore against the capillary pressure;
3) the globules coalesce to form a continuous oil phase, and the flow of oil will be comparable with (but not identical to) that of a single liquid in a material of smaller porosity.

Production of an oil well, we infer, takes place mainly in the third stage, if not entirely.

The question is, how does the effective permeability to oil change as the water saturation changes? We are interested in relative changes, so we express the effective permeability relative to the intrinsic permeability. The *relative permeability to oil* is therefore defined

$$k_{ro} = k_o/k \tag{9.1}$$

where k is the intrinsic permeability and k_o is the effective permeability to oil in the wetted material. $k_o = k$ when the material is 100% saturated with water *or oil.* It might seem that we should make the distinction between intrinsic permeability and the coefficient of permeability (or hydraulic conductivity), but experiments suggest that relative permeability is not sensitive to changes of viscosity or density (see Leverett, 1939). The reason for this apparent paradox is that the intrinsic permeability k is a property of the material only, independent of the fluid, and so can be determined at 100% saturation of either liquid.

From Darcy's law,

$$k = Q_w \eta_w l / A \rho_w g \Delta h_w = Q_o \eta_o l / A \rho_o g \Delta h_o \tag{9.2}$$

when the material is entirely saturated with either liquid. The analogous expression for effective permeability to oil when both immiscible liquids are present is

$$k_o = Q'_o \eta_o l / A \rho_o g \Delta h'_o . \tag{9.3}$$

Hence,

$$k_{ro} = k_o/k - Q'_o \Delta h_o / Q_o \Delta h'_o . \tag{9.4}$$

We would therefore expect relative permeability to be independent of viscosities and densities*.

* It has also been reported that relative permeability is independent of the specific discharge, $q = Q/A$. This is as one would expect, because these are in the nature of intrinsic permeabilities, which can be measured over a range of q provided the flow is within the realm of Darcy's law.

Reports that relative permeability varies with the pressure gradient arise, it seems, from the use of an incomplete form of Darcy's law (see Appendix, p. 207) and a failure to appreciate that k can be determined with any liquid.

They write

$$k = q_w \eta_w l / \Delta p$$

and

$$k_o = q'_o \eta_o l / \Delta p$$

so that

$$k_{ro} = k_o/k = q'_o / q_w \eta_w .$$

But q'_o is proportional to Δh_o and the larger the quantity Δh_o, the larger the quantity q'_o. But Δh_o does not appear in their formulation because it was cancelled out on the assumption that Δp was the same for both liquids. This is not necessarily so in an experiment using various pressure gradients on the *mixtures*.

Relative permeability is determined experimentally, and it is found that the relative permeability to oil (or any other immiscible, non-wetting liquid) reduces sensibly to zero before the oil saturation is reduced to zero – typically before it is reduced to 15%. There have been various explanations for this (see Bear, 1972, pp. 459-466, for a recent discussion of the whole subject), but clearly they are related to the second stage outlined earlier – the work required to force discrete globules of oil from one pore to another.

A point of fundamental importance must now be raised. We found in Chapter 3 that liquid flow through porous glass with very small passages obeyed Darcy's law, and therefore conclude that the static wetting layer may be very thin indeed, to be measured in molecules. This is in conflict with our inference that the wetting liquid, water, is static when the oil saturation is high. It seems likely, therefore, that the real situation is that most of the flow through porous material is in the central regions of the pores (as Lindquist, 1933, pp. 87-88, suggested) with trivial contribution from the peripheral pore space*. As the oil saturation decreases, so more and more water moves into the space of important flow. Once the oil loses physical continuity, more energy is required for its movement – and while oil globules remain, they impede water flow much as if there is a reduction of porosity (there is a reduction of effective porosity). There is a concomitant increase in tortuosity, because oil occupies the central space of the pores; and when there are discrete globules, they act as if they were small solid grains, so reducing the parameter d (the harmonic mean diameter of the matrix through which the liquid flows). *The presence of two immiscible liquids, or a liquid and a gas, reduces the effective permeability to both.*

We therefore approach relative permeability in a semi-quantitative manner through the porosity component of intrinsic permeability, assuming that the hydraulic radius term, $\{fd/(1-f)\}^2$, (as we have just inferred) and the shape factor C are independent of saturation while oil is flowing.

Let s be the water saturation, or proportion of the pore space occupied by water as the wetting liquid, then sf is the loss of effective porosity for oil due to the water saturation. From equation (3.18),

$$k \propto f^{1.5m-0.5} f^2/(1-f)^2, \qquad (9.5)$$

where m is the cementation factor that normally ranges from about 1.3 for unconsolidated sands to about 2 or a little more for consolidated sands.

For the analogous expression for effective permeability to oil, we note that the oil-water interface is not stationary while the oil is flowing, and so write

$$k_o \propto (f - sf)^{1.5m_o - 0.5} f^2/(1-f)^2 \qquad (9.6)$$

and the relative permeability to oil while the oil is flowing is

* *cf.* Versluys' (1919) concept of *washing* the salt water out in the transition zone of a Ghijben-Herzberg lens –a case of flow of two miscible liquids.

$$k_{ro} = k_o/k = \frac{(f - sf)^{1.5m_o - 0.5}}{f^{1.5m - 0.5}}, \tag{9.7}$$

where m_o is analogous to m, but is strictly a variable depending on the saturation s. For water flow, the oil is an effective reduction of porosity that causes an increase in tortuosity by denying to water the central spaces of the pores. For water saturations that allow the oil to flow, effective permeability to water will be approximated by

$$k_w \propto (sf)^{1.5m_w - 0.5} f^2/(1 - f)^2, \tag{9.8}$$

where m_w is larger than m_o for water saturations less than one (and greater than zero) on account of the greater tortuosity. Hence, the relative permeability to water is given by

$$k_{rw} = k_w/k = \frac{(sf)^{1.5m_w - 0.5}}{f^{1.5m - 0.5}}. \tag{9.9a}$$

When the water saturation is such that the oil ceases to be a continuous phase throughout the pore spaces, the oil not only reduces the effective porosity for water but also acts as a static boundary and so leads to an effective reduction in hydraulic radius. The harmonic mean diameter of the static components is also reduced. Ignoring changes in the harmonic mean diameter, we therefore write

$$k^*_{rw} \simeq k_{rw}\,[s(1 - f)/(1 - sf)]^2 \tag{9.9b}$$

We must now examine the quantities m, m_o and m_w. While m is a material constant determined from the Formation Resistivity Factor, m_o and m_w are variables that depend on the saturation. Each approaches m as its saturation approaches unity. Since oil flows in the central spaces of the pores, we can probably take m_o to be approximately equal to m without serious error for our purposes. By virtue of the water's exclusion from the central spaces of the pores, m_w will be larger than m – considerably larger when the water saturation is low. We can get an idea of the range from Leverett's (1939, p. 153, fig. 3) calibration curve for relative electrical conductivity versus saturation, on the assumption that relative electrical conductivity (C_r) is analogous to relative permeability to water. His data are closely approximated by $C_r = 1.1\, s^{2.2}$, but the exponent varies from about 1.5 as $s \rightarrow 1$ to about for $s = 0.2$. This suggests (by equating the exponent to $1.5m_w - 0.5$) that m_w in his unconsolidated sands ranged from about 1.3 for $s = 1$ (as it should be) to about 2 for $s = 0.2$ We therefore arbitrarily take $m_w = 2m - ms$, so that $m_w = m$ when $s = 1$ and $m_w \rightarrow 2m$ as $s \rightarrow 0$.

Introducing these substitutions into equations (9.7) and (9.9), we now write as our approximations to relative permeability:

$$k_{ro} = (1 - s)^{1.5m - 0.5}; \tag{9.10}$$

$$k_{rw} = \frac{(sf)^{1.5(2m - ms) - 0.5}}{f^{1.5m - 0.5}}; \tag{9.11a}$$

$$k^*_{rw} = k_{rw}\,[s(1-f)/(1-sf)]^2. \tag{9.11b}$$

Figure 9-3 is a plot of these hypothetical curves for $m = 1.6$ and $f = 0.3$, such as we might expect for a clean sandstone. The lower curve on the right is for k^*_{rw} when the 'irreducible' oil saturation has been reached. The dashed curve above the others is the sum of the two relative permeabilities while oil is flowing. Figure 9-4 is a plot of the family of curves for $m = 1.3$ and $m = 2.0$, the normal range, and porosities of 35% and 20% respectively.

The porosity hypothesis for relative permeability leads to curves very similar in shape and position to those found by experiment. The asymmetry is in the same sense as that found by experiment, with the curves crossing when the water saturation is more than 50% (see Leverett, 1939; Levorsen, 1967, p. 111, fig. 4-5; Bear, 1972, p. 460). There are several points of interest.

It has long been noted that the sum of the two relative permeabilities, $k_{ro} + k_{rw}$, is less than unity; and this has been attributed variously to differences of viscosity and to the entrapment of oil globules in 'dead-end' pores. It is sometimes referred to as

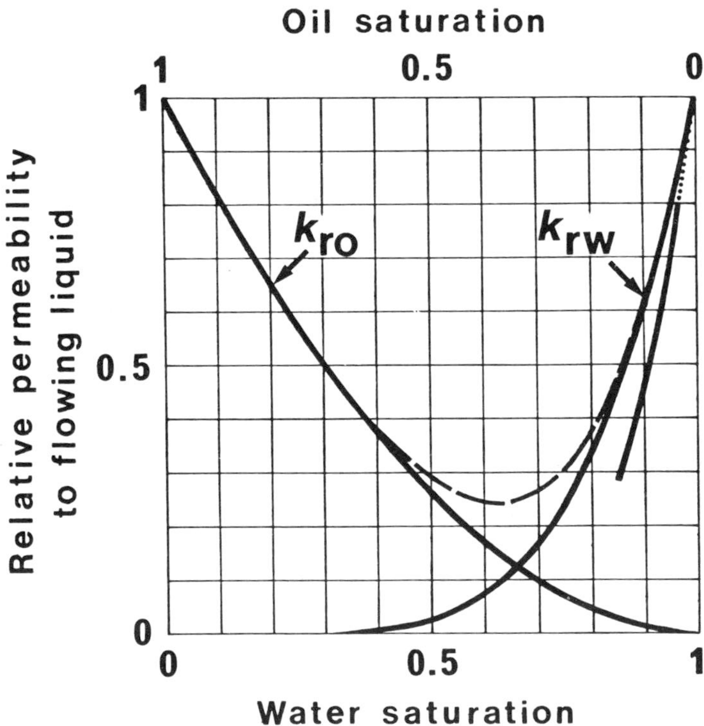

Fig. 9-3. Hypothetical relative permeabilities of two immiscible liquids plotted against saturation. Dashed line is the sum of k_{ro} and k_{rw}. Porosity 30%, $m = 1.6$. (*Note*: These curves should apply to any two immiscible liquids because the relationship is independent of liquid properties: for discussion, we have taken water to be the wetting liquid, oil to be the non-wetting liquid.)

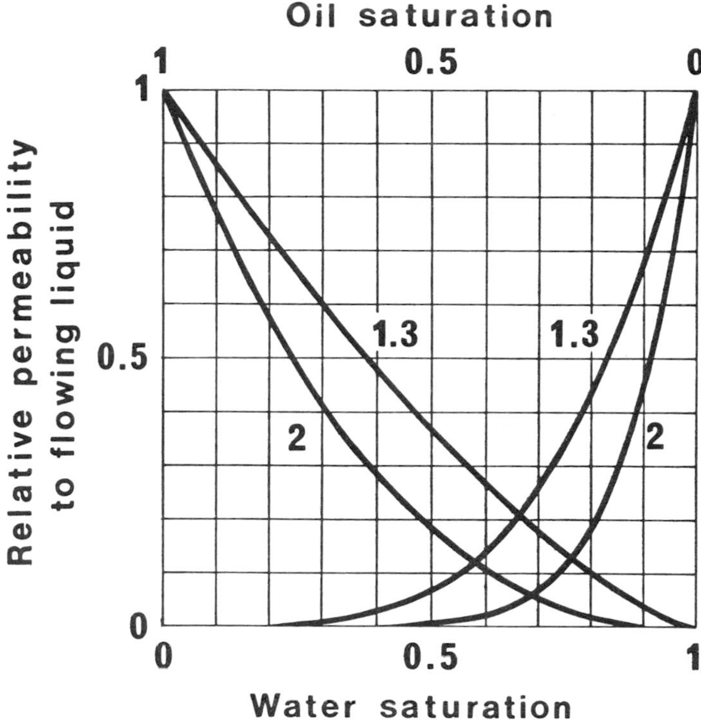

Fig. 9-4. Hypothetical relative permeability curves for $m = 1.3, f = 0.35$, and $m = 2.0$ and $f = 0.2$.

the *Jamin effect*. Figure 9-3 shows the sum of the hypothetical relative permeabilities plotted against saturation: this curve is also very similar in shape and position to those obtained experimentally. It is therefore interesting to note that our simplistic porosity hypothesis predicts for two-liquid flow *in homogeneous, isotropic* materials that the sum of the relative permeabilities should be less than unity – indeed, very much less than unity over most of the range of saturations – without recourse to any other effects.

The curves (both experimental and these) suggest that capillary action on the discontinuous oil phase is not *necessarily* important except for the highest water saturations, at least 75%. But we see why the 65% oil saturation level is the approximate economical and practical limit of oil-well production. The water-cut (proportion of water in the total being produced) is given by the ratio $k_{rw}/(k_{rw} + k_{ro})$, and the water/oil ratio by k_{rw}/k_{ro}. These are plotted as percentages in Figure 9-5 for the data of Figure 9-3.

Early production contains very little water, but soon after the water saturation reaches 35% the water-cut increases rapidly, and this salt water must be separated from the oil and disposed of. This effect is superimposed on the decline in total production (rapid decline in oil production) due to the reduction in the sum of both

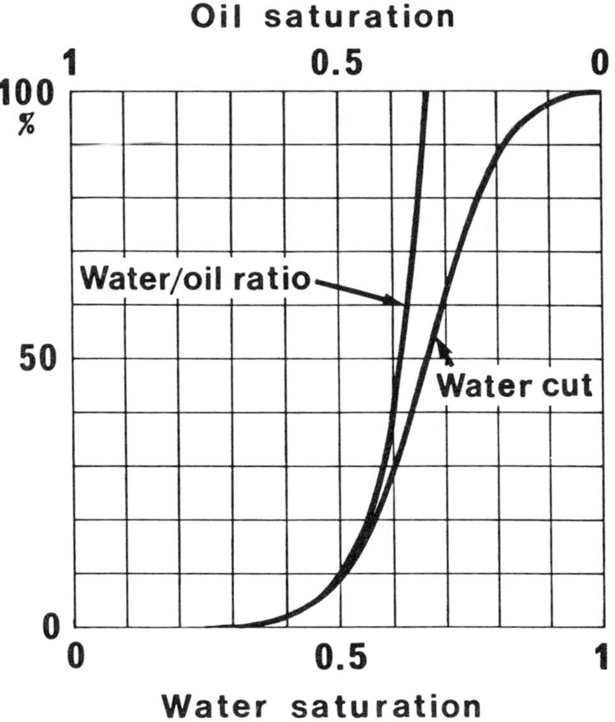

Fig. 9-5. Hypothetical proportions of water and oil produced at different saturations ($m = 1.6, f = 0.3$).

relative permeabilities. The 100% water production level is reached effectively before 100% water saturation in the pores is reached; but the model predicts that there will be a loss of relative permeability to water when the 'irreducible' oil saturation is reached. As this oil saturation is approached in real sediments, there will probably be a gradual shift to a point on the lower curve.

If a rock is oil-wet, then the pattern of relative permeabilities will not be the same, but rather like the mirror image. However, in this case the water is for practical purposes an unlimited resource, and only rapid production without pressure maintenance is likely to recover a significant proportion of the oil in place.

It will occur to the reader, as it has occurred to the oil companies, that significantly greater reserves of recoverable oil would be made available if a method could be found for recovering the residual oil. The term 'secondary recovery' applies to methods of reducing the amount of residual oil in final stages of normal production: 'tertiary recovery' for the recovery of the rest. Secondary recovery methods are mainly pressure maintenance by water injection into the reservoir outside the oil limits (and/or gas into the cap of the reservoir) with the object of sweeping the oil upwards into the structure as depletion proceeds and the oil is replaced by water. Tertiary recovery, still in the pilot-project stage with most companies, usually involves the injection of a solvent into the water, with the intention to taking oil into solution, and recovering it at the surface (and re-using the solvent). Our model does not suggest any practical alternative, unfortunately.

Interest in relative permeability should not be confined to petroleum reservoir engineers. If pore water migrating from abnormally-pressured shales contains gas in solution, the upward component of migration may take it to pressures at which the gas comes out of solution (Chapman, 1972; Hedberg, 1974). This will immediately reduce the effective permeability to water in a material of already low permeability (at first, down the lower line on the right of Figure 9-3, for example), and further expulsion of water will be impeded. Its pressure will therefore rise, and some or all of the gas may be taken back into solution – allowing the pore water to move again more freely, until the gas again comes out of solution. (If the gas coalesces to form a space of continuous gas phase, it is not clear what will happen to this gas unless its perimeter reaches a permeable bed, when it will be relatively easily expelled.) Whatever the final outcome of such a process may be, it is clear that upward migration of water will be seriously impeded, so enhancing the mechanical instabilities discussed in Chapter 7. Note, however, that the same tendency would appear to be absent from the downward-migrating pore water at the bottom of the shale unless the absolute pressure is decreasing. Perhaps this is why so many oil reservoirs seem to have their petroleum source rock stratigraphically higher in the sequence.

Infiltration of rain-water into soil and the underlying rocks is also a case of flow of two immiscible fluids, the second being air. Although this is a most complex phenomenon, with the air moving in the opposite direction to the water, the principles are the same as those we have discussed. It is a subject that has been intensively studied by many people for many years (see Bear, 1972, p. 439ff, for a recent detailed review and discussion).

Consider rain falling on a thin permeable soil that rests on a homogeneous, isotropic sand that is water wet. Between the surface and the water-table the pore spaces are, in general, partly filled with water (called *vadose* water), and partly with air. Once the water saturation exceeds some minimum level (the *specific retention*), water will flow downwards under gravity, displacing the air upwards. The relative permeability of the sand to water would follow a curve similar to those in Figure 9-4 on the right, but modified somewhat by the opposing air flow.

As a first approximation, we would suppose that soon after rain starts falling, the relative permeability of the sand to water will be greater near the surface than at some greater depth in the vadose zone. This will cause a 'front' of water to move downwards to the water-table. Once this front reaches the water-table, flow will tend to stabilize over the vadose zone, with a fixed proportion of the rain-water infiltrating into the sand. The water saturation in the vadose zone typically increases downwards to the water-table and with it, we suppose, the relative permeability to water until the point is reached that the air becomes a discontinuous phase, and the capillary pressure must be overcome to expel the remaining air. We are therefore forced, at last, to consider capillary pressure and surface tension.

Capillary pressure and surface tension

It is well known that if a glass tube of small internal diameter is inserted into a bowl of water (Fig. 9-6) water rises in the tube above the level in the bowl, and the shape of the air/water interface (the meniscus) is curved, as in Figure 9-7.

As regards the elevation of the water in the tube, it is evident that the air pressure is essentially the same in the capillary tube as on the surface of the water in the bowl; and since the pressure in the water at the level of the water surface in the bowl is zero (taking atmospheric pressure as datum, that is, gauge pressure), and the pressure in the water column in the tube decreases upwards, the pressure in the elevated water column must be negative, and there is a pressure differential at the meniscus. This interface therefore acts as if it were an elastic membrane in a state of tension, with the greater pressure on the concave side. This is the surface tension.

The amount of elevation, h_c, leads to the concept of capillary pressure:

$$p_c = -\rho g h_c = -p_w \tag{9.12}$$

The maximum amount of capillary rise is that at which the upward component of capillary force and the weight of the elevated water are equal, that is:

$$\rho g h_c \pi r^2 = \sigma 2\pi r \cos\theta$$

and

$$p_c = \frac{2\sigma}{r}\cos\theta; \quad h_c = \frac{2\sigma}{r\gamma}\cos\theta, \tag{9.13}$$

where σ is the surface tension, γ the specific weight of the water, r the radius of the capillary tube, and θ is the contact angle (Fig. 9-7) ($\cos\theta \simeq 1$ for water on quartz).

For *pendular water* around the point of contact of two spheres, it can also be shown that the capillary pressure is related to the two radii of curvature on the surface of the water:

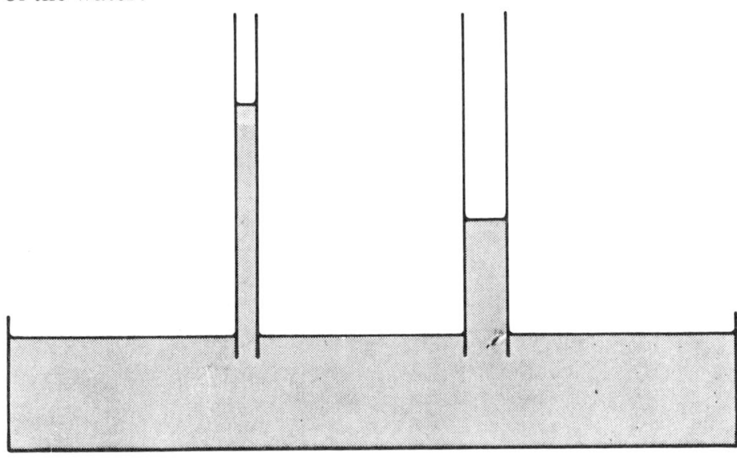

Fig. 9-6.

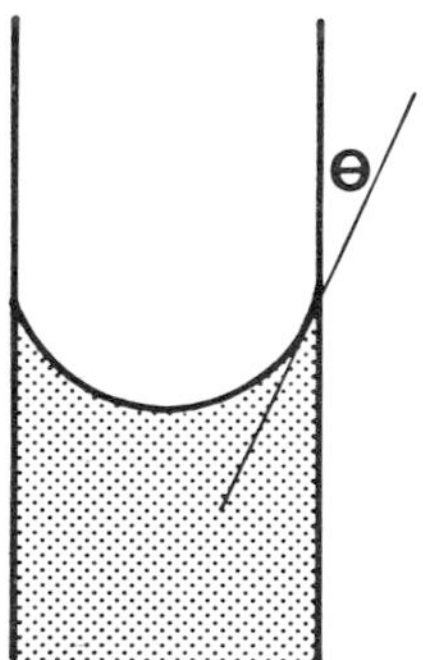

Fig. 9-7.

$$p_c = \sigma\left(\frac{1}{r'} + \frac{1}{r''}\right) = 2\sigma/r_* \qquad (9.14)$$

where r_* is the harmonic mean of the two radii, taken to be negative since p_c is negative). σ has dimensions MT^{-2}: it is defined in terms of work per unit of area. Surface tension is known as *interfacial tension* between two liquids.

Sediments are not measurable in these terms. The contact angle θ is only constant when the fluids are static: it is greater for an advancing water surface than for a retreating surface (examine a dew drop on a petal). It varies with water quality, as does surface tension, and the nature of the solid. But the phenomenon of capillary rise of water in sediment has long been known (Perrault, 1674, reported experiments), and the capillary rise and capillary pressure are inversely proportional to a dimension r_* that can be regarded as a measure of pore size. We have seen that intrinsic permeability (dimensions L^2) contains a measure of pore size (the hydraulic radius) and so would expect that

$$h_c \propto (1 - f)/fd, \qquad (9.15)$$

where d is, as before, the harmonic mean grain diameter.

The amount of capillary rise (known qualitatively as the *capillary fringe*) therefore varies according to the nature as well as the texture of the sediment. It amounts to less than one metre for sands, to a few metres for sediments of small hydraulic radius. (A similar effect is present at the oil/water contact in an oil field that bounds an oil accumulation: the transition is not at a surface, but over a zone – thin in oil reservoirs – in which the saturations change.)

We therefore see that the water-table should be defined not as the surface of 100% water saturation, but the surface in the water at which the pressure is atmospheric. The pores are 100% saturated for a short distance above this level, and above that, the water saturation decreases.

It therefore seems unlikely that capillary forces are important in water infiltration through a water-wet sand much above the level of 100% saturation, unless the water-table is near the land surface. The infiltration rate will be determined by the minimum relative permeability of the sand to water in the profile near the surface, and any rainfall in excess of that amount will become surface run-off. This minimum relative permeability in the profile will remain less than unity while air remains to be expelled.

In reality, of course, there are many complicating variables involved, such as irregular rainfall, dissolution of air in water, evaporation and condensation of water in the vadose zone – but these are variations on the theme.

It has not been my purpose in this chapter to present a detailed analysis of the phenomena discussed, but rather to illustrate through these phenomena two important applications of some of the concepts developed in earlier chapters. Further study of two-phase flow can begin with the references listed, and the general references at the end of Chapter 1. Such applications, though, are not unrewarding.

SELECTED BIBLIOGRAPHY

Bear, J., 1972. *Dynamics of fluids in porous media.* American Elsevier, New York, London, and Amsterdam, 764 pp.

Bear, J., Zaslavsky, D., and Irmay, S., 1968. *Physical principles of water percolation and seepage.* UNESCO, Paris, 465 pp.

Bridgeman, M., 1969. Recent advances and new thresholds in petroleum technology. (8th Cadman Memorial Lecture.) *J. Inst. Petroleum,* 55: 131-140.

Chapman, R.E., 1972. Primary migration of petroleum from clay source rocks. *Bull. American Ass. Petroleum Geologists,* 56 (11): 2185-2191.

Fowler, W.A., Boyd, W.A., Marshall, S.W., and Myers, R.L., 1971. Abnormal pressures in Midland Field, Louisiana. *In*: Houston Geol. Soc., *Abnormal subsurface pressure: a Study Group report* 1960-1971. Houston Geol. Soc., Houston, pp. 48-77.

Hedberg, H.D., 1974. Relation of methane generation to undercompacted shales, shale diapirs, and mud volcanoes. *Bull. American Ass. Petroleum Geologists,* 58 (4): 661-673.

Hubbert, M.K., 1951. Mechanical basis for certain familiar geologic structures. *Bull. Geol. Soc. America,* 62: 355-372.

Hubbert, M.K., and Willis, D.G., 1957. Mechanics of hydraulic fracturing. *Trans. American Inst. Mining Metallurgical Petroleum Engineers,* 210: 153-166. (Discussion: 167-168.)

Leverett, M.C., 1939. Flow of oil-water mixtures through unconsolidated sands. *Trans. American Inst. Mining Metallurigcal Engineers* (Petroleum Division), 132: 149-169. (Discussion: 169-171.)

Levorsen, A.I., 1967. *Geology of petroleum* (2nd edition). W.H. Freeman & Co., San Francisco, 724 pp.

Lindquist, E., 1933. On the flow of water through porous soil. 1er *Congrès des Grands Barrages* (Stockholm, 1933), 5: 81-101.

Perrault, P., 1674. *De l'origine des fontaines.* Pierre le Petit, Paris, 353 pp.

Rose, W., 1957a. Fluid flow in petroleum reservoirs. I. – The Kozeny paradox. *Illinois State Geol. Surv. Circular* 236 (8 pp.).

Rose, W., 1957b. Fluid flow in petroleum reservoirs. II. – Predicted effects of sand consolidation. *Illinois State Geol. Surv. Circular* 242 (14 pp.).

Versluys, J., 1916. *De capillaire werkingen in den bodem.* (Thesis, Technische Hoogeschool, Delft, The Netherlands.) W. Versluys, Amsterdam, 136 pp.

Versluys, J., 1919. De duinwater-theorie. *Water,* 3 (5): 47-51.

Versluys, J., 1931. Can absence of edge-water encroachment in certain oil fields be ascribed to capillarity? *Bull. American Ass. Petroleum Geologists,* 15 (2): 189-200.

APPENDIX. REVIEW OF COMMONLY-QUOTED WORKS ON DARCY'S LAW

Our purpose here is to review, concisely, the main results of commonly-quoted authors of the last 60 years or so, confining ourselves to those who sought to develop the theme, rather than use the work of others*. Their expressions have been translated into the notation of this book.

Blake (1923), working in the context of chemical engineering, was apparently the first to use dimensionless groups for the plotting of experimental permeability data. His groups lead to the following form of Darcy's law:

$$q = C \frac{f^3}{\eta S^2} \frac{\Delta p}{l} . \tag{A.1}$$

Although he was concerned with both air and water, his groups are strictly valid only for gas or horizontal flow of liquid, and there is no useful way in which his work can be modified for the earth sciences. He did, however, recognize that q is a notional velocity and substituted for it the term q/f; and he introduced the concept of hydraulic radius, though not by name, by substituting f/S for the grain diameter.

Kozeny (1927b) worked in the context of soils and irrigation. His work is of central importance, not only because it was analytical and thorough, but also because his results have been incorporated into both chemical ingineering and earth sciences. Much of the paper is devoted to capillary movements, and is peripheral to our theme.

He first obtained an analytical expression of Darcy's law (without, however, quoting him directly: he does not seem to have been aware of Blake's work). He begins the development with the equation

$$q = K \cdot I = K \frac{\Delta h}{l}$$

where K 'die Durchlässigkeit und Δh die auf dem Stromlinienstück l verbrauchte Druckhöhe darstellt' – 'where K is the permeability and Δh the loss of pressure head

* It is regretted that works published in languages other than English, German, and French, have not been consulted because translations cannot be relied upon if the translator does not have an intimate knowledge of the subject.

over the stream length l' (quoted with change of notation). It is unfortunate that he used the word *Drukhöhe* because he clearly understood the nature of h (Kozeny, 1927a, p. 67). We shall take I to be the hydraulic gradient.

Kozeny (1927b) recognized that q is a notional velocity, and inserted correcting factors to write his equation (20),

$$q = \frac{\rho g}{\eta} I C \frac{l}{l_t} \frac{f^3}{S^2}, \tag{A.2a}$$

where C is a shape factor of restricted range of values, S is the specific surface, and I is the gradient of total head. He continued his equation (20) (our (A.2a)) for the case of spherical grains,

$$q = \ldots = \frac{\rho g}{\eta} I C \frac{l}{l_t} \frac{f^3}{36\,(1-f)^2} d^2. \tag{A.2b}$$

Accepting I as the hydraulic gradient, Kozeny's equation (20) (our (A.2a) and (A.2b)) is essentially correct, although the role of tortuosity has not been taken fully into account.

Kozeny also examined the meaning to be attached to the mean diameter and that of the fraction between two sieves, and suggested the harmonic mean and two variants (op. cit., pp. 305-306).

He perceived the effect of porosity and wrote (op. cit., pp. 277-278)

$$K = K_1 \frac{f^3}{(1-f)^2} \tag{A.3}$$

having shown that experimental data supported the porosity factor $f^3/(1-f)^2$ better than others that had been proposed *even for grains of irregular shape*.

Fair and Hatch (1933), working in the context of hydraulic engineering and sand filters, used dimensional analysis of pipe flow as the basis, and so also the concept of hydraulic radius. They considered the meaning of d and concluded that there is some shape factor that, when multiplied by $1/d$, will give the correct specific surface. They used the geometric mean of adjacent sieve sizes, $d_n = \sqrt{d'd''}$, for the mean of the fraction on the assumption of a 'geometric size distribution'. They derived the expression (op. cit., p. 1558, equation (3))

$$\frac{\Delta h}{l} = \frac{c_1}{g} \frac{\eta}{\rho} \frac{(1-f)^2}{f^3} q \left[\frac{c_2}{W} \sum_{i=1}^{n} \frac{w_i}{d_i} \right]^2 \tag{A.4a}$$

which can be re-arranged

$$q = C \frac{f^3 d^2}{(1-f)^2} \frac{\rho}{\eta} g \frac{\Delta h}{l}, \tag{A.4b}$$

where C combines $1/c_1$ (they found c_1 to have a value of about 5) and the geometric shape factor $1/c_2^2$ (where c_2 has a limited range of values from 6 for spheres to about 7.7): and d is the weighted harmonic mean grain diameter.

This expression differs from that of our equation (3.15a) only in neglecting the tortuosity factor, and the relatively trivial difference of mean for the sieve fraction.

Wyckoff, Botset, Muskat, and Reed (1933, 1934), working in the context of petroleum reservoir engineering, proposed to define a unit of permeability and call it a 'darcy' (although the basis of the unit had been suggested by Nutting, 1930); and they described the apparatus to measure permeability. (The papers cover the same ground: the 1934 paper is the main one, to which we shall refer.)

They defined the permeability of a porous medium as 'the volume of a fluid of unit viscosity passing through a unit cross section of the material under a unit pressure gradient in unit time', and gave an incomplete form of Darcy's law

$$q = \frac{k}{\eta} \frac{\Delta p}{l} \quad \text{(dimensionally correct)} \tag{A.5a}$$

which they develop into

$$k = \frac{\eta q}{\Delta p / l} = \frac{\eta Q l}{A(p_1 - p_2)}. \tag{A.5b}$$

They proposed a unit of permeability to be called '"darcy" after D'Arcy' to be equal to

$$\frac{\eta \times \text{cm}^3 \times \text{cm}}{\text{sec} \times \text{cm}^2 \times \text{atmospheres}} = \frac{\eta \times \text{cm}^2}{\text{sec} \times \text{atmos}} = \frac{\text{poises} \times \text{cm}^4}{\text{sec} \times \text{dynes}} \tag{A.6}$$

with dimensions $[L^2]$, to be measured only within the realm of Darcy's law (they regarded turbulence as setting in as Darcy's law fails at the upper limit).

Although they wrote 'The rate of flow of a fluid through any system is dependent upon two fundamental quantities: the pressure gradients applied, and the resistance to flow developed along the channel' (Wyckoff et al., 1934. p. 161), and wrote their expression for Darcy's law accordingly, it seems that they were actually measuring the difference of total head in their apparatus. In the description of the apparatus, they emphasized (op, cit., pp. 171-172) the use of manometers rather than gauges 'with the zero pressure points properly referred to the zero-flow levels in the columns'. The sample holder is vertical in their apparatus.

Carman (1937, 1938, 1939). Carman's papers are in the context of chemical engineering (1937, 1938), the results of which were offered to agricultural science (1939).

He started with an incomplete statement of 'd'Arcy's law', modified Kozeny's results, and wrote (1937, p. 152, equation (8))

$$q = \frac{fR^2}{c_1\eta}\, g\, \frac{\Delta p}{l}\left(\frac{l}{l_t}\right)^2 \tag{A.7a}$$

and, making the substitution $R = f/S$, wrote what he called Kozeny's equation

$$q = \frac{f^3}{c_1\eta S^2}\, g\, \frac{\Delta p}{l}\left(\frac{l}{l_t}\right)^2. \tag{A.7b}$$

This is not what Kozeny wrote, nor is it dimensionally correct.

He then reviewed the experimental work and concluded that the value of c_1 is constant at about 5 (as Fair and Hatch, 1933, had found), so the surface area of powders could thus be estimated since all the other quantities in the expression are measurable.

Were it not for the influence Carman exterted on subsequent work, we would not pursue the matter further: Kozeny's original expression is better. As Carman stated in his introduction (1937), his purpose was to emphasize 'the importance of the method of plotting by dimensionless groups introduced by Blake', and to use permeability methods to determine 'the surface of powders': he introduced nothing new.

In his 1938 paper, Carman may have stated Darcy's law correctly when he referred to 'head loss', but before the end of the page, pressure gradient has re-appeared. He modified 'Kozeny's equation' to the 'more convenient form'

$$q = \frac{f^3}{c_1\eta S^2}\, \rho\, g\, \frac{\Delta h}{l}. \tag{A.8}$$

He then claimed that the dimensionless hydraulic gradient, $\Delta h/l$, is equal to $\Delta p/\rho l$. This implies that pressure has the dimensions ML^{-2}.

These difficulties notwithstanding, he obtained good agreement with experimental results.

Hubbert (1940, 1956/1957) examined Darcy's law from an earth-science point of view, specifically ground-water hydrology, as part of a broader study. The relevant results may be summarized:

a) liquids move down an energy gradient, the energy per unit of mass at a point in the liquid being defined as its potential

$$\Phi_{\text{liq}} = gh = g\left(\frac{p}{\rho g} + z\right). \tag{A.9}$$

(His discussion and derivation were more general, applying to any fluid: Hubbert, 1940, p. 802.)

b) Darcy's law can be expanded and generalized

$$q = K\frac{\Delta h}{l} = N d^2 \frac{\rho}{\eta} g \frac{\Delta h}{l} = N d^2 \frac{\rho}{\eta} \frac{\Delta \Phi}{l}, \tag{A.10}$$

where N is a dimensionless shape factor, and $Nd^2 = k$, the intrinsic permeability (op. cit., p. 816).

c) He wrote 'the' Reynolds number applicable to flow through porous media $q\rho d/n$ (op. cit., p. 811), and
d) he pointed out that the unit of permeability, the darcy, was defined on the basis of an erroneous expression of Darcy's law (equation (A.5) above) (op. cit., pp. 921-922).

Hubbert's 1957 paper, written by invitation for the celebration of Darcy's centenary, examined in detail Darcy's experiments, their physical meaning, and the constitution of K. The expressions he derived (in the present context) are essentially those of his 1940 paper, but he elaborated on the shape factor N (Hubbert, 1957, p. 42-46) introducing a porosity factor $f^3/(1-f)^2$ (op. cit., p. 44, equation (63); and p. 45, equation (73)). (See Note in Selected Bibliography.)

Krumbein and Monk (1942) looked at intrinsic permeability as a function of the size parameters of unconsolidated sand from the sedimentologists' point of view. Using Hubbert's (1940) work as a basis they derived the expression 'for size distributions that satisfy the normal logarithmic probability law' (Krumbein and Monk, 1942, pp. 9-10)

$$k = Nd^2 = N' d^2 e^{-a\sigma_\phi} = (k/GM\xi^2)_0 \, GM\xi^2 \, e^{-a\sigma_\phi}, \tag{A.11}$$

where k is the intrinsic permeability, N is a shape factor, $GM\xi$ is the geometric mean diameter of the particles, σ_ϕ is the log standard deviation. The constants $(k/GM\xi^2)_0$ and a are determined experimentally.

It should be noted that porosity was kept constant close to 40%, and is not included explicitly in their equation. It should also be noted that the range of permeabilities studied, 57 to 1,555 *darcies*, exceeds the usual range of sediment permeabilities by about three orders of magnitude.

Rose (1945a, 1945b) took a dimensional approach in the engineering context, but does not seem to have reached any new conclusions of interest to geologists. He did consider the upper limit of Darcy's law, plotting his own and other workers' data as a Reynolds number against a coefficient of resistance, and found a departure from linearity at a Reynolds number of about 10. He regarded this (1945a) as being the onset of turbulence.

Schneebeli (1955) reported experiments conducted in the hydraulic-engineering context near the upper limit of Darcy's law, using the dimensionless groups

$$\left(\frac{gd\Delta h}{q^2 l}\right) \text{ and } \left(\frac{qd\rho}{\eta}\right). \tag{A.12a}$$

The first was called the coefficient of resistance (*frottement*): like Rose's, it is the reciprocal of a Froude number. The second is, of course, a Reynolds number. From these groups, the following expression can be derived for the realm of Darcy's law,

$$q = C\, d^2 \frac{\rho}{\eta} g \frac{\Delta h}{l} \tag{A.12b}$$

In Figure 3-9 we plotted Schriever's (1930) data against these groups. It is clear that the porosity term is required.

The main importance of Schneebeli's work is his confirmation of the fact that Darcy's law begins to fail well before the onset of turbulence. Darcy's law failed with spherical material at a Reynolds number ($qd\rho/\eta$) of about 5, while turbulence began at $Re = 60$. With angular fragments, Darcy's law failed at $Re = 2$, with turbulence at about 60, as with spherical material. The limiting Reynolds number for Darcy's law found by Schneebeli is substantially the same as that found by Lindquist (1933), who also concluded that this was well before the onset of turbulence (op. cit., p. 91).

DISCUSSION

This brief review of selected, but commonly-quoted literature dealing with the flow of fluids through porous media indicates a substantial consensus in the approach and the results. Recognition that q is a notional velocity dates back further than the period reviewed, at least to Slichter (1899). It was early recognized that there is a component of permeability due to the material alone, and that the surface area of solids bounding the flow is important among the quantities affecting resistance to flow. Unfortunately, the seeds of an error were also sown early.

Darcy himself used the word *pression*, but he qualified it quite unambiguously. Slichter (1899, p. 329) wrote Darcy's law as a function of pressure difference, although he recognized the role of gravity (op. cit., p. 331, equation (7)) and introduced a potential function. But the confusion persisted because on the next page he wrote of 'equipotential or equipressural surface' as though the two were synonymous. Not all the subsequent papers using pressure alone were in error in the context in which they were written: the error is introducted when expressions intended for gas, or the horizontal flow of liquids, are applied to non-horizontal flow of liquids. In many, but not all, earth science applications, no significant error is introduced by assuming horizontal flow.

It is a tragedy of science that Darcy's name should be linked to the unit of permeability with such an imperfect definition – the more so since the proposers were probably measuring what Darcy himself measured. If Wyckoff et al. (1933, 1934) had gone from equation (A.5a) to $q = (k/\eta)\rho g(\Delta h/l)$ and so to

$k = Q\eta l / A\rho g\,(h_1 - h_2)$, they would have defined a unit of permeability of greater value.

Had earth scientists been content with the papers of Darcy and of Fair and Hatch (both written in an hydraulic engineering context, as was Darcy's, but one sufficiently close by analogy to ground-water and petroleum geology) or those of Kozeny, the topic might have been better understood. Much of the difficulty stems from Carman's papers – but Carman is not entirely to blame for this. He was writing for chemical engineers; and if they find his expression useful and wish to call it the Carman-Kozeny equation, that is their affair. Nevertheless, one would expect engineers to use expressions that are dimensionally homogeneous.

The choice of 'mean grain diameter' as the characteristic dimension by all but Kozeny and Fair and Hatch has obvious disadvantages in that it is only indirectly related to the 'mean pore diameter' through which the liquid flows. If Krumbein and Monk (1942) had taken d to be the weighted harmonic mean grain diameter, they would have found that the greater the variance, the smaller the harmonic mean is relative to the geometric mean. In their samples 10 to 20 (op. cit., p. 5, table 1), all with a geometric mean diameter of 1 mm, they would have found a steadily decreasing harmonic mean as the variance increased, and it is possible that the observed decrease in permeability with increasing variance would have been accounted for. Their equation seems to involve an empirical correction to the choice of central tendency for statistical description of a sedimentary rock.

It seems that most of those who have used Krumbein and Monk's equation have failed to notice that porosity is not included, and that the porosity for which the equation was derived was about 40%.

There is no doubt that Hubbert's treatment of Darcy's law (see Hubbert, 1969) is rigorous and useful, nor is there any doubt that lumping the various dimensionless quantities into a single shape factor N has practical advantages. But the porosity factors determined earlier by Kozeny (1927) and Fair and Hatch (1933) have their advantages also.

Kozeny's and Fair and Hatch's recognition of the nature of the grain diameter was also a significant step that has largely been neglected. It may be argued that sedimentologists have become so settled on the median and the geometric mean as their measures of central tendency that the harmonic mean is a trivial refinement of no practical use. This is hardly so.

There is, unfortunately, no means of converting one measure of central tendency to another: but the harmonic mean grain size of a sediment is as easily computed from the mechanical analysis as the geometric mean – it is merely a different operation on the same numbers.

SELECTED BIBLIOGRAPHY

Beard, D.C., and Weyl, P.K., 1973. Influence of texture on porosity and permeability of unconsolidated sand. *Bull. American Ass. Petroleum Geologists*, 57 (2): 349-369.

Blake, F.C., 1923. The resistance of packing to fluid flow. *Trans. American Inst. Chemical Engineers*, 14 (for 1922): 415-421.

Burke, S.P., and Plummer, W.B., 1928. Gas flow through packed columns. *Industrial and Engineering Chemistry*, 20 (11): 1196-1200.

Carman, P.C., 1937. Fluid flow through granular beds. *Trans. Instn Chemical Engineers*, 15: 150-166.

Carman, P.C., 1938. The determination of the specific surface of powders. *J. Soc. Chemical Industry*, 17: 225-234.

Carman, P.C., 1939. Permeability of saturated sands, soils and clays. *J. Agricultural Science*, 29 (2): 262–273.

Corrsin, S., 1955. A measure of the area of a homogeneous random surface in space. *Quarterly of Applied Mathematics*, 12 (4): 404–408.

De Ridder, N.A., and Wit, K.E., 1965. A comparative study on the hydraulic conductivity of unconsolidated sediments. *J. Hydrology*, 3: 180-206.

Fair, G.M., and Hatch, L.P., 1933. Fundamental factors governing the stream-line flow of water through sand. *J. American Water Works Ass.*, 25 (11): 1551-1565.

Fraser, H.J., 1935. Experimental study of the porosity and permeability of clastic sediments. *J. Geology*, 43 (8, part 1): 910-1010.

Graton, L.C., and Fraser, H.J., 1935. Systematic packing of spheres – with particular relation to porosity and permeability. *J. Geology*, 43 (8, part 1): 785-909.

Green, W.H., and Ampt, G.A., 1911. Studies on soil physics. Part I. – The flow of air and water through soils. *J. Agricultural Science*, 4 (1): 1-24.

Green, H., and Ampt, G.A., 1912. Studies on soil physics. Part II. – The permeability of an ideal soil to air and water. *J. Agricultural Science*, 5 (1): 1-26.

Hubbert, M.K., 1940. The theory of ground-water motion. *J. Geology*, 48 (8): 785-944.

Hubbert, M.K., 1956. Darcy's law and the field equations of the flow of underground fluids. *Trans. American Inst. Mining Metallurgical Petroleum Engineers*, 207:222-239. (Also in: *J. Petroleum Technology*, v. 8, October 1956.)

Hubbert, M.K., 1957. Darcy's law and the field equations of the flow of underground fluids. *Bulletin de l'Association Internationale d'Hydrologie Scientifique*, no. 5: 24-59. (*Note*: in this Appendix, reference is made principally to Hubbert, 1957, rather than 1956. The papers are the same, but 1957 will be found in Hubbert, 1969, together with Hubbert, 1940, and Darcy, 1856.)

Hubbert, M.K., 1963. Are we retrogressing in science? *Bull. Geol. Soc. America*, 74: 365-378.

Hubbert, M.K., 1969. *The theory of ground-water motion and related papers.* Hafner Publ. Co., New York and London, 310 pp.

Kozeny, J., 1927a. Über Grundwasserbewegung. *Wasserkraft und Wasserwirtschaft*, 22: 67-70, 86-88, 103-104, 120-122, 146-148.

Kozeny, J., 1927b. Über kapillare Leitung des Wassers im Boden (Aufstieg, Versickerung und Anwendung auf die Bewässerung). *Sitzungsberichte der Akademie der Wissenschaften in Wien*, Abt. IIa, 136: 271-306.

Krumbein, W.C., and Monk, G.D., 1942. Permeability as a function of the size parameters of unconsolidated sand. *Technical Publs American Inst. Mining Metallurgical Engineers*, 1492 (11 pp.)

Lindquist, E., 1933. On the flow of water through porous soil. *Premier Congrès des Grands Barrages* (Stockholm, 1933), 5: 81-101.

Morcom, A.R., 1946. Flow through granular materials. *Trans. Instn Chemical Engineers*, 24: 30-36. (Discussion: 36-43.)

Nutting, P.G., 1930. Physical analysis of oil sands. *Bull. American Ass. Petroleum Geologists*, 14: 1337-1349.

Pryor, W.A., 1973. Permeability-porosity patterns and variations in some Holocene sand bodies. *Bull. American Ass. Petroleum Geologists*, 57 (1): 162-189.

Rose, H.E., 1945a. An investigation into the laws of flow of fluids through beds of granular materials. *Proc. Instn Mechanical Engineers*, 153: 141-148.

Rose, H.E., 1945b. On the resistance coefficient-Reynolds number relationship for fluid flow through a bed of granular material. *Proc. Instn Mechanical Engineers*, 153: 154-168.

Rose, H.E., and Rizk, A.M.A., 1949. Further researches in fluid flow through beds of granular material. *Proc. Instn Mechanical Engineers*, 160: 493-511.

Rose, W.D., 1959. Calculations based on the Kozeny-Carman theory. *J. Geophysical Research*, 64 (1): 103-110.

Schneebeli, G., 1955. Expériences sur la limite de validité de la loi de Darcy et l'apparition de la turbulence dans un écoulement de filtration. *La Houille Blanche*, 10 (2): 141-149.

Schriever, W., 1930. Law of flow for the passage of a gas-free liquid through a spherical-grain sand. *Trans. American Inst. Mining Metallurgical Engineers* (Petroleum Division), 86: 329-336.

Slichter, C.S., 1899. Theoretical investigation of the motion of ground waters. *Annual Report U.S. Geol. Survey* (1897-98), 19 (2): 295-384.

Wyckoff, R.D., Botset, H.G., and Muskat, M., 1932. Flow of liquids through porous media under the action of gravity. *Physics*, 3: 90-113.

Wyckoff, R.D., Botset, H.G., Muskat, M., and Reed, D.W., 1933. The measurement of the permeability of porous media for homogeneous fluids. *Review Scientific Instruments*, new series, 4: 394-405.

Wyckoff, R.D., Botset, H.G., Muskat, M., and Reed, D.W., 1934. Measurement of permeability of porous media. *Bull. American Ass. Petroleum Geologists*, 18 (2): 161-190.

CONVERSIONS

PREFIXES FOR SI UNITS

10^{-18}	atto	a	
10^{-15}	femto	f	
10^{-12}	pico	p	
10^{-9}	nano	n	
10^{-6}	micro	μ	
10^{-3}	milli	m	
10^{-2}	centi	c	Further use discouraged
10^{-1}	deci	d	Further use discouraged
10	deka	da	Further use discouraged
10^{2}	hecto	h	Further use discouraged
10^{3}	kilo	k	
10^{6}	mega	M	
10^{9}	giga	G	
10^{12}	tera	T	

AREA

$1\ \text{M}^2 = 1.55 \times 10^3$ square inches
$= 10.764$ square feet
$= 1.196$ square yards
$= 2.4711 \times 10^{-4}$ acres

DENSITY

$1\ \text{kg m}^{-3} = 10^{-3}\ \text{g/cm}^3$
$= 6.2428 \times 10^{-2}$ pounds/cubic foot
$= 1.9420 \times 10^{-3}$ slugs/cubic foot

FLOW RATE

$1\ m^3\ s^{-1}$ = 35.31445 cubic feet/second (cusec)
= 1.3198×10^4 gallons/minute (GPM), British or Imperial
= 1.5850×10^4 U.S. GPM
= 7.9189×10^5 Imp. gallons/hour (GPH)
= 9.5102×10^5 U.S. GPH
= 1.9005×10^7 Imp. gallons/day (GPD)
= 2.2825×10^7 U.S. GPD

FORCE

1 newton (N) = $1\ kg\ m\ s^{-2}$
= 10^5 dyne
= 0.1020 kgf
= 0.2248 lbf
= 7.233 poundals

GRADIENT

$1\ N\ m^{-1}$ = 6.8519×10^{-2} lbs/ft
$1\ kgf\ m^{-1}$ = $9.80665\ N\ m^{-1}$
= 0.6720 lb/ft
$1\ Pa\ m^{-1}$ = $1\ N\ m^{-2}\ m^{-1}$
= $0.1020\ kgf\ m^{-2}\ m^{-1}$
= $1.0197 \times 10^{-5}\ kgf\ cm^{-2}\ m^{-1}$
= 4.4208×10^{-5} psi/ft
1 °C/km = 0.55 °F/1000 ft

LENGTH

1 m = 39.37 inches
= 3.2808 feet
= 1.0936 yards
= 0.5468 fathoms
= 6.2137×10^{-4} miles
= 10^6 microns (μ)
= 10^{10} angstrom (Å)

PERMEABILITY (intrinsic)

1 darcy = 10^{-8} cm^2 (for practical purposes)
1 md = 10^{-11} cm^2

PRESSURE

1 pascal (Pa) = 1 N m^{-2}
= 1×10^{-5} bar
= 1.0197×10^{-5} kg cm^2
= 9.8687×10^{-6} atmospheres
= 10 dynes/cm^2
= 1.4504×10^{-4} pounds/sq. inch (psi)
= 2.0886×10^{-2} pounds/sq. foot

TIME

1 second = 1.1574×10^{-5} days
= 1.6534×10^{-6} weeks
= 3.8023×10^{-7} months (30.44 days)
= 3.1688×10^{-8} years (365.25 days)

VELOCITY

1 m s^{-1} = 3.6 km/hour
= 3.2808 feet/second
= 2.2369 miles per hour
= 1.9426 knots

VISCOSITY

1 Pa s = 1 N m^{-2} s = 10 poise = 10^3 cP
= 0.6720 pounds/foot · second (lb mass)
= 2.0886×10^{-2} lbf · sec/sq. ft
1 m^2 s^{-1} = 10^4 stokes = 10^6 cSt
= 1.5500×10^3 square inches/second
= 10.7639 square feet/second

VOLUME

1 m^3 = 219.969 Imperial gallons
= 264.173 U.S. gallons
= 8.107×10^{-4} acre-feet
= 6.2898 barrels (bbl) of oil

WEIGHT (see also FORCE)

1 kg weight = 2.2046 pounds weight

GLOSSARY

AQUICLUDE. A relatively impermeable rock unit that confines an *aquifer* (above, below, or both).

AQUIFER. A porous and permeable, or fissured, rock unit that contains exploitable water. Originally used for any water-bearing stratum, but probably meant what it now means. Norton (1897, p. 130) revived 'a term of Arago's': but Arago (1835, p. 229) uses the normal French adjective 'aquifère' for water-bearing.

Arago, D.F.L. 1835. Sur les puits forés, connus sous le nom de puits artésiens, de fontaines artésiennes, ou de fontaines jaillissantes. *Annuaire, Bureau des Longitudes*, 1835: 181-258.

Norton, W.H., 1897. Artesian wells of Iowa. *Iowa Geological survey*, 6: 113-428.

AQUITARD. A rock unit that is not as permeable as an *aquifer* nor as impermeable as an *aquiclude*.

ARTESIAN. Wells or aquifers in which the energy of the water is sufficient to raise it above the land surface without the use of pumps are said to be artesian. An aquifer is artesian if its total head is greater than the elevation of the land above the same datum.

Some would define it in terms of local water-table rather than land surface, but compare *subartesian*.

Name derived from *Artesium*, latin name for *Artois*, historical province of northern France, where early artesian wells were constructed.

CAPILLARY FRINGE, RISE. The uppermost part of the *zone of saturation* above the level *in* the water that is at atmospheric pressure; or the zone in which the water is at less than atmospheric pressure; or the zone of 100% water saturation above the *water-table* (*sensu stricto*). The c. rise is the thickness of the c. fringe.

COEFFICIENT OF PERMEABILITY. The coefficient K in Darcy's law when written $q = K\Delta h/l$. It is related to *intrinsic permeability* by $K = k\rho g/\eta$. Dimensions LT^{-1}. Synonym: *hydraulic conductivity*.

DARCY. Unit of permeability based on incomplete expression of Darcy's law (see Appendix, p. 207). Common unit, millidarcy (md). Dimensions L^2.

DYNE. Unit of force in c.g.s. system. It is that force that gives to a mass of 1 gramme an acceleration of 1 cm per second per second. Dimensions MLT^{-2}.

ELASTIC STORAGE. See *Storage coefficient*.

EQUIPOTENTIAL SURFACE, LINE. A surface or line on which points of equal fluid potential lie. Not synonymous with *potentiometric surface*. Synonym: *isopoten-*

tial, which is not to be preferred, but is too commonly used to be dropped now.

FIELD CAPACITY. The water remaining in *soil* above the *capillary fringe* after extended period of drainage, expressed as percentage by weight of dry weight of soil. Cf. *Specific retention*, *Pellicular water*. Dimensionless.

FROUDE NUMBER. Ratio of the products of two lengths and accelerations, $L_1 L_1 T_1^{-2}/L_2 L_2 T_2^{-2}$. In general, it is the number V^2/dg, where V is a velocity, d a characteristic length, and g the acceleration due to gravity. It is interpreted as the ratio of inertial to gravitational forces. In engineering practice, the Froude number is usually taken as the square root of the number given above, i.e., $V/\sqrt{dg}$.

GEOSTATIC. Pressures and pressure gradients due to the gravitational load of the total overburden are called geostatic. Undesirable synonym: *lithostatic*.

GROUND WATER. In practice, term restricted to usable subsurface water, fresh or brackish.

HEAD. Refers to the vertical length of a column of fluid, usually liquid: ambiguous when not qualified. It is an energy per unit of weight, with dimensions of *Length*, not pressure.

Elevation head (less desirably, *potential head*): the potential energy per unit weight due to elevation, i.e., the elevation of the point at which pressure p is measured, above (+) or below (−) an arbitrary horizontal datum. $h_e = z$.

Pressure head: the vertical length of a column of liquid supported, or capable of being supported, by pressure p at a point in that liquid. $h_p = p/\rho g$.

Velocity head: head due to the kinetic energy of the liquid. Negligibly small in most geological contexts. $h_v = V^2/2g$.

Total head: the algebraic sum of the pressure and elevation heads (neglecting velocity head).

Static head: pressure head in a borehole that is not producing, cf. *static level*.

HYDRAULIC CONDUCTIVITY. The coefficient K in Darcy's law when written $q = k\Delta h/l$. It is related to *intrinsic permeability* by $K = k\rho g/\eta$. Dimensions LT^{-1}. Synonym: *Coefficient of permeability*.

HYDRAULIC RADIUS. A measure of the size of the passage for fluids when the shape is irregular. It is the area of transverse section of liquid divided by the *wetted* perimeter, or equally the volume of liquid divided by the wetted surface area. Dimension L.

HYDRAULIC GRADIENT. The difference of *total head* (see *Head*) divided by the macroscopic length of *porous material* between the points where the total head is measured – properly, the steepest gradient at a point in the fluid. Note that the gradient of a *potentiometric surface* is not strictly the hydraulic gradient unless the aquifer is horizontal (but it may be a sufficiently close approximation). Hydraulic gradient is not the same thing as pressure gradient. Dimensionless.

INTERFACIAL TENSION. When two immiscible fluids are in contact in a capillary tube or fine-grained porous material, the interface between the two liquids acts as if it were an elastic membrane in a state of tension. Cf. *surface tension*. Dimensions MT^{-2}.

INTRINSIC PERMEABILITY. The component of permeability that is ascribable to the porous material alone, independent of the physical properties of the fluid passing through it. Dimensions L^2.

ISOPIESTIC LINE. Line of equal potential or total head. Synonym of *equipotential*, which is to be preferred for the same reasons that *potentiometric* is to be preferred over *piezometric*.

ISOPOTENTIAL SURFACE, LINE. A surface or line on which points of equal fluid potential lie. Not synonymous with *potentiometric surface*. Synonym: *equipotential*, which will be preferred by those who do not mix greek and latin roots.

ISOPYCNIC SURFACE, LINE. A surface or line of equal density.

LITHOSTATIC. Undesirable synonym of *geostatic* and overburden (adj.) in context of pressures and pressure gradients.

PELLICULAR WATER. Water in the *vadose zone* that does not move because it adheres to the solid surfaces. Cf. *field capacity*.

PHI (ϕ) UNITS. Grade scale in sedimentology. $\phi = -\log_2$ (diameter in mm).

PHREATIC WATER. Water in the *zone of saturation*, i.e., it is now synonymous with *ground water*.

PIED. Old French measure of length. 1 toise = 6 pieds = 1.949 m, so 1 pied = 0.325 m. Piés were presumably about the same.

See letter to The Times, May 2, 1903, *in:* Gregory, K., (Ed.), 1978. *The first cuckoo: letters to The Times since* 1900. Unwin Paperbacks, London, p. 45.

PIEZOMETRIC SURFACE. See *potentiometric surface*.

POISE. Unit of dynamic or absolute viscosity in c.g.s. system. A force of one gram will maintain unit rate of shear between two surfaces of unit area unit distance apart when the viscosity of the liquid between the surfaces is one poise (P). Units are: dyne-second per square centimetre ≡ gram per centimetre second ≡ poise. Dimensions $ML^{-1}T^{-1}$.

POROSITY. A rock is said to have porosity if there is space between the solid grains. If space is due to fractures, it is called fracture porosity. Quantitatively, it is the volume of void space divided by the total volume of solids and void space, $f = v_p/(v_p + v_s)$. Usually spoken of as a percentage. Dimensionless. Related to *void ratio* by $f = \varepsilon/(\varepsilon + 1)$.

POTENTIOMETRIC SURFACE. The surface obtained by mapping the total head of an aquifer. The potentiometric surface can be contoured (*equipotential lines*) and fluid flow is in the direction normal to these contours, in the direction of decreasing total head or potential. The synonym *piezometric surface* is not to be preferred because it is a matter of potential, not pressure.

PRESSURE. Force per unit of area. The pressure at a point in a fluid is the force per unit area acting on an element of volume that is small in relation to the bulk of the liquid, but large in relation to the molecules. SI Units: pascal (Pa) = newtons per square metre (N m^{-2}) = kg m^{-1} s^{-2}. Dimensions $ML^{-1}T^{-2}$.

RETAINED WATER. Water in the *saturated zone* that will not drain by gravity (cf. *field capacity*, *specific yield*).

REYNOLDS NUMBER. Ratio of the products of two lengths and velocities, $L_1L_1T_1^{-1} / L_2L_2T_2^{-1}$. In general, it is the number Vd/v, where V is a velocity, d is a characteristic length, and v is the kinematic viscosity. It is interpreted as the ratio of inertial to viscous forces acting on the system.

RUNOFF. Rain water that flows on the surface of the ground to streams, rivers, etc, without passing through porous and permeable soil or aquifers. River runoff, or discharge, however, includes the ground-water component. Stormwater runoff is synonymous with rain water runoff because this usually has the quality of a storm since light rain leads to evaporation and infiltration, not runoff, unless the water table is already at the surface.

SATURATION. The saturation of a rock with respect to a fluid is the proportion of the pore space filled with that fluid. Dimensionless.

SLUG. Unit of MASS in American usage, as distinct from the pound, which is the unit of weight or force. A slug is the unit of mass that acquires an acceleration of 1 foot per second per second when acted upon by a force of one pound weight. From Newton's second law of motion, *acceleration* = *force*/*mass*. Thus, unit acceleration is acquired by 32.174 lb mass, which equals one slug. This is usually rounded to 32.2 lb mass. For conversions, 1 slug = 14.594 kg mass.

SPECIFIC RETENTION. The proportion of soil or rock occupied by water adhering to solid surface after gravity drainage. Dimensionless (volume/volume). Cf. *specific yield*.

SPECIFIC STORAGE. See *storage coefficient*.

SPECIFIC SURFACE. Surface area of solids in bulk volume of porous material. Dimensions L^{-1}.

SPECIFIC VOLUME. The volume of unit mass of substance, reciprocal of mass density. Dimensions $M^{-1}L^3$.

SPECIFIC WEIGHT. Weight divided by volume of substance or material, ρg. Dimensions $ML^{-2}T^{-2}$.

SPECIFIC YIELD. Volume of water that will drain by gravity from saturated soil or rock, divided by the gross volume of the soil or rock. What is left behind is *retained water*, hence *specific retention*. Cf. *storage coefficient*, *storativity*.

STATIC LEVEL. The level of the water surface in a well when it is not producing and has stabilized. This level may be given relative to the ground surface or the bottom of the well; but if given relative to some *horizontal* datum surface becomes synonymous with *total head*.

STOKE. Unit of kinematic viscosity, which is the absolute or dynamic viscosity divided by the mass density of the fluid, η/ρ. Units: square centimetres per second. Common practical unit, centistoke (cSt). Dimensions L^2T^{-1}.

STORAGE COEFFICIENT, STORATIVITY. When the total head of an aquifer changes, the volume of water in the aquifer changes. The storage coefficient is the volume of water released from (or taken into) storage in unit volume of the aquifer per unit decline (increase) in total head. It consists of two parts: one related to the compressibility of the water, the other related to the compressibility of the rocks skeleton (*elastic storage*). For an unconfined aquifer, *storage coefficient* = *specific yield*. Dimensions $L^3/L^4 = L^{-1}$.

It is also defined in terms of a vertical column through the aquifer, with unit basal area. In this case, it is dimensionless.

SUBARTESIAN. Said of wells or aquifers in which the energy of the water is sufficient to raise it above the water-table, but not above the ground surface. Cf. *artesian*. There is merit in calling all wells in an artesian aquifer artesian, but there is merit also in distinguishing those that need pumping by calling them subartesian.

SURFACE TENSION. In a capillary tube, or fine-grained porous material, the air/water interface acts as if it were an elastic membrane in a state of tension, supporting a pressure differential across it. This is surface tension. But it is defined in terms of the work required to separate unit area of interface. Cf. *interfacial tension*. Dimensions MT^{-2}.

TRANSMISSIVITY. For horizontal flow in an aquifer, the product of *hydraulic conductivity* and the thickness of the aquifer. Dimensions $\mathrm{LT^{-1}L = L^2T^{-1}}$.

VADOSE WATER, ZONE. Water in the *zone of aeration* descending to the water table. (Derivation variously ascribed to latin *vado* = go, walk, rush, and *vadosus* = shallow.)

VOID RATIO. Engineers prefer void ratio to *porosity*. The void ratio is the volume of pores divided by the volume of solids. It is related to porosity by $\varepsilon = f/(1-f)$.

WATER-TABLE. In practice, the free water surface of an unconfined aquifer in a well. Strictly, it is the level *in* the water in the *zone of saturation* at which the pressure is atmospheric, i.e., the base of the *capillary fringe*.

WEIGHT DENSITY. The weight of a substance or material divided by its volume, i.e., ρg. Synonym: *specific weight*. Dimensions $ML^{-2}T^{-2}$.

WET, WETTED, WETTING. When two immiscible fluids saturate a porous material, one preferentially adheres to the solid surfaces, excluding the other. The adhering fluid is the *wetting* fluid or wetting phase, the other is the non-wetting fluid or phase. Sands are usually *water-wet*.

ZONE OF AERATION. The subsurface zone above the *zone of saturation* in which the voids are partly filled with air, partly with water and water vapour.

ZONE OF SATURATION. The subsurface zone below the level at which the voids are entirely filled with water at hydrostatic pressure, i.e., strictly, below the capillary fringe.

POSTSCRIPT

POROSITY ESTIMATION FROM SONIC LOGS

Work done while this book was in press indicates that fractional porosity of mudstones (with no carbonate) can be *estimated* from the formula

$$f = f_0\left(\frac{\Delta t_{sh} - \Delta t_{ma}}{\Delta t_0 - \Delta t_{ma}}\right)$$

where Δt_0 is the transit time when porosity is f_0 (both obtained from plots of Δt_{sh} and porosity against depth, extrapolated to $z = 0$), Δt_{ma} is the transit time when $f = 0$ – the so-called matrix transit time.

Reasonable values of the material constants are: $f_0 = 0.5$; $\Delta t_0 = 165$ μs/foot, 540 μs/metre; $\Delta t_{ma} = 55$ μs/ft, 180 μs/m. The formula therefore reduces to

$$f = (\Delta t_{sh} - 55)/220 \text{ for sonic logs in μs/foot;}$$

$$f = (\Delta t_{sh} - 180)/720 \text{ for sonic logs in μs/metre.}$$

Pore-water expulsion by compaction from porosity f_1 to f_2 can then be estimated, as a proportion of bulk volume of mudstone at porosity f_2, from

$$\hat{q} = (f_1 - f_2)/(1 - f_1)$$

(see Chapman, 1972; reference on p. 204).

For sandstones and carbonates, the fractional intergranular porosity can be estimated from the formula

$$f = 1 - (\Delta t_{ma}/\Delta t)^x \approx 1 - (\Delta t_{ma}/\Delta t)^{1/2}.$$

This formula results from the assumption that the sonic path is through solids only, and that porosity affects the length of this path (a sort of tortuosity in solids). It is quite different in form and in theory from those in use at present, but it gives good results and can be used when there is no core data. The following values of Δt_{ma} are suggested: sandstones, 55μs/ft, 180 μs/m; carbonates, $47\frac{1}{2}$ μs/ft, 155 μs/m. Local experience may indicate other values.

It is a simple matter to make scales so that porosity can be read directly from the sonic log.

SUBJECT INDEX

AUTHOR INDEX